SECRETS - NEVER HEARD UNTIL NOW - OF THE BOOK OF REVELATION

JOHN ZACHARY

Harvard House
PUBLISHING

SECRETS- NEVER HEARD UNTIL NOW - OF THE BOOK OF REVELATION

Finding unexpected strength when the world tramples on your faith!

John Zachary

Secrets - *never heard until now* - of the Book of Revelation

Bible quotations are from the New American Standard Bible (NASB) unless otherwise specified when quoted.

Cover Design by www.delaney-designs.com
Editor: Tom Wall

ISBN: 978-1-951885-01-4

Harvard House
PUBLISHING

DEDICATIONS

I dedicate this book to Jim M., the original supporter of my ministry. I had revealed the details in chapters one and six to Jim. He decided to teach the research to others. While studying the materials, he passed away with that book in his hands. He had discovered details that reveal we can expect to live forever!

I also dedicate this book to my parents because they instilled faith in all their children. They appear to have added to this book in a way that was beyond their control. I recorded this inspirational mystery in chapter four.

I also dedicate this book series to Pastor Charlie Garrett. You will learn why as you read the Expect to Live Forever book series.

CONTENTS

PREFACE

Across two millennia, the book of Revelation seems to have lain in a deep crypt, shrouded by its riveting mysteries. At times, some Christian leaders declared it should not be in the Bible. Have you been waiting for the day that the Holy Spirit would shock the world and validate its mysterious texts. Would you dare to read the deep secrets hidden in this Inspirational Christian Mystery?

Does your life have meaning? Are your friends losing their faith? Do you fear death? Open your heart and mind to the reality of the spiritual realm. *Discover why you should expect to live forever*!

1

THE ULTIMATE SECRET

Do you keep secrets from your friends?

In general, people keep secrets to keep goodwill, or so everyone can relax. However, should you keep a personal mystery that has the potential to affect every person living on the Earth?

I ask that question because my mysteries are from spiritual sources. I am writing about thousands of sacred experiences. These secrets and experiences began more than four decades ago. One of these deep mysteries may upset many people. Regardless of this potential issue, I have made the decision to let people know. Because I believe it is in the best interest of everyone to know these secrets. Then you can consider them.

I am writing about a mysterious experience related to the book of Revelation. It is the last book of the Bible. This information links our time to the first-century author of the book of Revelation. What I am about to disclose begins to reveal a profound reality about each of us. The evidence shows why you can expect to live forever!

There are two sides to my deep secrets. First, I chose to keep my spiritual experiences to myself for three decades. I have had many thousands of these experiences. During those years, I was busy studying the Bible in-depth. I used science to find the precise dates of foretold events. This was based on the only "time-oriented" prophecy in the entire world. This time-based prophecy foretells of future events to an exact day. In fact, the prophetic texts are precise. This evidence shows that many people were guided by a Spiritual Being, whom we refer to as God.

It appears that the guided people did not know of this guidance. However, the result of this guiding action caused specific events to happen on an exact day. The biblical texts foretell the precise date.

Between the experiences I write about and my research, you will begin to gain insight into yourself. You will realize that you have a soul, which means "*You will live forever!*"

This memoir will prepare you for the future. Since the book of Revelation predicts global-wide events, can you fathom change that will affect every culture and person on planet Earth?

The second side to my secrets is from studying biblical prophecy. I released this information to the public around twenty-five years ago. However, there were some issues in the research that I did not know how to defend. Most of the research was correct. At that time, I could not find persuading evidence to validate the study against the critics that refused to accept these cutting edge materials.

I am sure that most people would have given up on it. My idea is to confirm spiritual things using a science-based method. Consider why this is a valid approach. If we can verify that scriptures did foretell the future to an exact day, that is not normal[1]. This would strongly suggest that the Source of those words knows the future.

Humans cannot foretell the future many centuries ahead of time. For example, do you know what events will occur on an exact date in the year 2501? Say on July 28?

Finding such words, then seeing how precise they are, makes the spiritual realm credible. Such words would be from a Source that exists beyond the control of time. Also, this Source exists beyond the control of space. The Source would live in the past, the present, and the future.

If the Source exists outside time, then foretelling the exact date of a specific future event would be easy. In fact, the Source would know every detail about you before you were ever born,[2] even before the universe began.

It is most important to ask, "*Why didn't I just quit doing this research due to its difficulty?*"

This is where spiritual experiences caused me to go on. It did not matter if it was hard to do and took time. The spiritual event I am about to reveal would, in fact, support that there must be an answer.

Before I disclose this experience, I believe that you should know something. I made the decision that before I share these experiences, to get counsel. I chose a conservative pastor, because I do not want to scare you or make you believe I am crazy. When I got the pastoral counseling, I expected my pastor to say, "*Do not write!*" However, to my surprise, that pastor encouraged me to write the materials.

After I began writing, I decided to assemble a group of people to criticize what I write. I attempted to find Christian people that would not approve of my spiritual experiences. However, no one discouraged me.

I believe it to be vital that you know that I looked for reasons to not publish these materials. My Christian friends seem to truly enjoy these materials. Are you ready to read an "Inspirational Christian Mystery?"

I KEPT GOING ON BECAUSE OF AN OUT OF BODY EXPERIENCE[3]. THIS happened to me on February 1, 1990. The following is a brief overview of this experience.

I write the details of this spiritual event in chapters six and seven.

I awoke at 2:30 am on February 1, 1990. I was fully awake as I lifted my head up off the couch where I was sleeping. My hands were tingling with the fiery presence of the Holy Spirit[4]. Then a thought went through my mind. That thought sounded like my voice. It would have seemed as if I was talking to myself. The words were: "***You are one of the prophets of the book of Revelation!***"

Now be assured, if that were all that happened, then I would have disregarded the entire sentence. For sure, you would not be reading this book. However, at the end of the final phrase, "***. . . of the book of Revelation***," I felt my spiritual arms and hands rise out of my physical body. The soul of my being became very obvious with this out of body experience. This out of body thing only happened to my hands and arms. The physical and spiritual parts of me stayed attached at my shoulders. Since you may be thinking, "*How would he know about his spiritual arms and hands? For sure, I have never personally felt my soul or anything of this nature!*"

The answer is simple. I knew and felt my spiritual arms and hands in the same way that we feel our physical arms and hands. However, I felt two distinct parts of my being. I felt my physical arms and hands. At the same time, I felt my spiritual arms and hands. Both the physical parts and the spiritual parts were separate and distinct during this out of body experience. In fact, I would expect that when you die, you will be fully aware of your soul!

The graphic below illustrates the fleshly arms and hands as well as the spiritual parts during this experience. Pay close attention to my hands. You will notice that my physical hands did not move. They remained in a position pointing toward each other.

In contrast, my spiritual hands moved to where they pointed straight up, giving praise to God. Then they stayed in that position for almost two-seconds. The grayscale part of this graphic stands for my spiritual parts, literally, my soul.

Please understand, this means that you, too, have a soul.

Spiritual Arms and Hands Contrasted with Fleshly Appendages

THE INSTANT MY SPIRITUAL ARMS AND HANDS STARTED TO RISE OUT OF my physical parts, a new thing happened. A new word flowed through my thought pattern. Also, the voice changed. It was not my voice. Instead, the thought pattern became a voice of praise. The new word was an extended "Whew!" This word took about two and a half seconds from start to finish.

The praise word began as my spiritual arms and hands rose above my physical arms. Then the voice continued to the maximum height above my physical body. Also, that voice was from the Spiritual Being, who created the universe and authored the book of Revelation. For sure, the voice was not my voice.

You can read about the details of the out of body experience[5] in chapters six and seven. For now, I must make an important point about what this spiritual experience reveals about the book of Revelation.

Is the Book of Revelation Credible?

Most biblical scholars think that the Apostle John wrote the book of Revelation. The composition occurred from AD 90 to 96. This viewpoint came from the early church fathers. They wrote about John during the second century.

However, as centuries passed, opinions on the book of Revelation changed. Church leaders began to question who authored the book. Some church leaders rejected the book of Revelation. They took the position that the book of Revelation should not be in the Bible. This meant that the book of Revelation is not an approved Canonical book. Why would this happen?

Perhaps this happened as the book did not make sense during the times in which those people lived. For sure, people deem the book of Revelation to be cryptic.

Now think about a Christian having an out of body experience about the book of Revelation. This Christian lives in the twenty-first century. What would this mean for the book of Revelation?

I feel that this certifies it is Canonical. Also, the book of Revelation (the apocalypse) has credible links to many other books in the Bible[6].

The author of the Book of Revelation

I believe that people who live closer to the event know more than those who live centuries later. Those early church fathers had more insight. I would expect that they knew the Apostle John was the author. Perhaps people talked to one another about the Apostle John as the author. However, they did not record those "secret conversations" on paper.

Why would people become wary of the book of Revelation?

It seems that texts in the book of Revelation did not fit their era. In contrast, the symbolism of this book is well-honed for the twenty-first century. In fact, the words are more relevant than all the previous centuries combined.

Consider that the Apostle John knew Jesus in person as a youth. John is one of three disciples in the inner circle of Jesus[7]. This is based on the New Testament story of the Transfiguration[8].

We know that John was a close follower of Jesus. They walked together, talked together, and ate together. They knew each other better than anyone else did. John was an eyewitness of Jesus and stood at the cross with Mary. In contrast, the other disciples fled from that scene. John saw the miracles, the crucifixion, and Jesus after he rose from the dead. The eyewitnesses of Jesus wrote clearly that he is the Messiah.

Maybe you think that the Bible is make-believe. Was it all made up?

Think about a follower of Jesus who has an out of body experience about the book of Revelation. This person is ordinary. For sure, he is no better than anyone else. Nevertheless, an out of body experience enhances his relationship with God.

Through this out of body experience, this person gets the message that he is one of the prophets of the book of Revelation! What does this reveal about the book of Revelation? Is it make-believe?

The Apostle John wrote the book of Revelation in the first century. Now we have a report of an out of body experience about the book of Revelation in the twenty-first century!

This strongly suggests there is a spiritual reality. Because two people, separated by 1,900 years, received spiritual words related to the book of Revelation!

'Separated by nineteen centuries' strongly suggests that the Source of the book of Revelation is supernatural. Time has no control over the Source of the revealed texts.

This out of body experience boosts the credibility of the book of Revelation. Imagine that John knew Jesus in person in the first century and he wrote the book of Revelation. Now a Christian living in the twenty-first century receives a spiritual message about the book of Revelation. Since the book of Revelation is about the future, could it be that we are living at the time of fulfillment?

Some people teach the book of Revelation is unimportant for our time. They explain that the events in this book took place during the Roman Empire[9]. However, an out of body experience in the twentieth century suggests that view is wrong. This experience indicates the book of Revelation will come alive and affect you. Are you willing to consider what this means for you?

Also, the book of Revelation is a book about Christian views of God. The book of Revelation confirms the Trinitarian view of God with the recorded words of Jesus.

> "*...saying, 'Do not be afraid; I am the first and the last, and the living One; and I was dead, and behold I am alive forevermore, and I have the keys of death and of Hades*" (Rev. 1:17-18)[10].

The book of Revelation uses symbolism to foretell the future. The symbols fit into the twenty-first century, like clay pressed into a mold. Sometimes the ideas link directly into the use of satellites. For example, the texts in Rev. 11:9-11 foretell that all people on planet Earth would be able to see newsworthy events at the same time. The period recorded 1,900 years ago is an event with a time limit of three and a half days. This would include the time for people to send gifts to one another. I have electronically mailed funds overseas in as little as fifteen minutes. How can an author living 1,900 years ago describe our world so well?

Everyone on Earth can now watch global news events within hours of such events. This was not possible until recent decades. Hence the book of Revelation has insightful prophetic visions about the twenty-first century. Such an exact report of our time described millennia ago supports that the words are from beyond space-time. The Source of these texts is Jesus Christ.

Be sure the book of Revelation will affect your life no matter what you believe or think about the Bible. I hope and pray that you are ready to learn and that you will choose to receive spiritual truth. Ask yourself, "*Am I ready for a transformation to obtain eternal life in paradise for the most fabulous party of all time?*"

Spiritual Report

THIS MEMOIR IS A REPORT OF WHAT I HAVE EXPERIENCED. ABOUT twenty-seven centuries ago, Isaiah the prophet recorded,

> "*Who has believed our report?*" (Isa. 53:1 NJKV).

I am merely following him. I am reporting to you about the reality of the spiritual realm. You can decide to accept or reject what I have written.

I have chosen to report these revelations after many decades have passed. It may be hard for you to digest that what I am saying to you are indeed real events. After you read this, you will understand that the events happening to me are something that was meant for you to learn. The genuineness of these experiences is my reason to reveal these spiritually driven events to you now. I must report based on; *"to whom much is given, much will be required"* (Luke 12:48 DARBY).

Will you begin to conceive that by reading the Expect to Live Forever book series, you will also have much to share with others. I hope you choose to do this with the motive of pure love.

This memoir reveals true stories that give insight into you. Will you will gain the vision that will help awaken you to the fact[11] that ultimate reality is spiritual?

Every human that will ever exist, in truth, is an eternal being. Because every human reflects the image of God (Gen. 1:27). I believe that your awakening will occur, especially if you read and study all my research in several books. For sure, that is your choice to make. I can only share what I have experienced with you here.

Conclusions

What can we conclude from this out of body experience with words spoken about the book of Revelation?

For sure, we know that there is a spiritual realm. I felt my spiritual arms and hands lifted out of my physical body. There is more to our being than our physical bodies. Finally, this serves to validate the book of Revelation is a Canonical book of the Bible. Moreover, this experience verifies that the Bible is the Word of God.

If you were present in the room during this experience, you would have seen me awake with my head lifted off the couch when this event occurred. However, it is unlikely you would have seen my spiritual arms and hands move. This may seem menial to you, but each chapter will reveal the objectivity of each experience. In fact, many of these experiences have objectivity. This means that you would perceive there is a spiritual realm.

PLEASE NOTE THAT THIS FIRST CHAPTER MAY NEED CLARIFICATION FOR some readers. The author has always accepted the Canonicity of the book of Revelation. In general, spiritual experiences do not verify the Canonicity of the Bible. However, spiritual experiences support, even validate, a spiritual realm.

Perhaps you feel like the author brags based on the materials in chapter one. However, by the time you have read chapter nine, you will discover that the author takes no credit for the information in the entire Expect to Live Forever book series.

2

AWE-INSPIRING AND ELEGANT WORDS

In chapter one, I revealed that I have two secrets. The second secret is related to my scientific study of prophecy. There is only one "time-oriented" prophecy in the entire world. All other predictions with a time-stamps to future events have failed.

The "key" to gaining insight into the spiritual realm is based on the following logic. We know that no human can foretell of a specific future incident that must happen on an exact date. This includes recording the texts on paper centuries before. What would it mean to find passages that predict a specific event on an exact date, centuries into the future?

It means that those unique words are from a Source that knows the future. For sure, we would decide to study the holy writ of the prophetic texts. We would conclude that the Source is a Supreme Being[1] that wants humans to know of His existence as well as other spiritual things.

In contrast, is it possible to use human reason or human feelings to understand God?

Humans cannot reason or feel what God is or should be. The human view of God would include "human error." In contrast, the word revelation entails the method that God uses to reveal spiritual truth[2]. In the Bible, this method involves a person "anointed" as a prophet. To this person, God reveals heavenly knowledge. This can include the foretelling of future events that unveil spiritual truth.

This Supreme Being reveals the future as well as information about the plans and purposes of our existence. God certainly wants us to study the recorded texts of the Bible. That is how we learn spiritual truth and begin to perceive that we will live forever.

Maybe you have concluded that you cannot trust the Bible. Consider that the materials in this chapter and the next books are based on logic, historical evidence, and biblical texts. If you choose to ignore what I have written, that is your choice. For sure, you have freedom of expression.

The "Time-Oriented" Prophecy

I SPENT YEARS RESEARCHING THE "TIME-ORIENTED" PROPHECY. I learned how to pinpoint the exact dates of historical events. To be successful, you must know the day of the week upon which an ancient incident occurred. To uncover this information required me to search through lots of evidence. I became a detective at work on unearthing the spiritual realm.

At this point, I will give you an overview of my logical approach. If you want more information, please read book two in the Expect to Live Forever book series, Eternal Life - *Why you should expect to live forever*[3]!

WE HAVE FOUND FOUR SETS OF 14,000 DAYS HIDDEN IN THE BIBLE.

These periods foretell the future. They also reveal spiritual truth about you. Would you like to know, without question, that you will live forever?

These 14,000-day periods are now events from the past[4]. Each of these occurred at a crossroads. This bulleted list includes what happened on the 14,000th day.

1. When Moses and Joshua lived: The 14,000th day is the precise day that they entered the Promised Land. This hidden set of 14,000 days turns out to be mystical.
2. When the prophets lived, who saw Solomon's Temple destroyed. Foreign soldiers burned down Solomon's Temple on the 14,000th day. This 14,000 days period sets up the New Testament. We will discover that New Testament events occur on the exact dates of the Old Testament events based on the Hebrew calendar.
3. When Jesus entered Jerusalem on Palm Sunday as the Messiah: Exactly 14,000 days into the future from Palm Sunday, Roman soldiers burned down Herod's Temple. The start and end dates repeat the same precise dates written into the Old Testament. We begin to recognize that both Solomon's Temple and Herod's Temple burned down on the 14,000th day. WHY?
4. In the twentieth and twenty-first centuries, when Jesus' prophetic words happened. The 14,000th day occurred on the Feast of Trumpets. That feast is symbolic of the Second Coming of Christ.

Consider an essential thing on this final bullet. Jesus foretold of these events, which occurred in the modern era. Jesus' words may be unveiling that the book of Revelation is about to take place. If this happens, it will affect every person on Earth. Here are two reasons why you should examine this.

1. I reported the out of body experience about the book of Revelation. I only made a report so you can consider this information.
2. The fact of repeating 14,000 days strongly suggests massive change is about to take place. Because spiritual knowledge comes from the final 14,000th day. The most recent period points to the Second Coming of Christ.

Evidence of God

This chapter gives a quick overview of the repeating 14,000 days cycles. These periods are uncovered by evidence from ancient history with eyewitness accounts. Also, what people wrote aligns with the positions of the Sun, the Moon, and the Earth. So, the exact science of astronomy backs up their statements.

However, I believe it is vital that you understand why the research supports the fact of a Creator. Specifically, the God of the Bible created the universe.

On the previous pages, I revealed when the 14,000 days cycles occurred. On the 14,000th day, we find a spiritual crossroads. What happens on the final day reveals spiritual truth.

If you want to gain insight, I suggest that you read one of the following books in the Expect to Live Forever book series.

- **Book 2:** Eternal Life - *Why you should expect to live forever*! If you do not like math and charts, this book is easy to read. Also, it is in an audio format.
- **Book 3:** Beyond - *The Coming Prince by Sir Robert Anderson*! This book has the math-based research, charts, ancient astronomy, logic, and many appendices.

Texts that Foretell the Future

You may ask. "*How do I know these passages actually predict the destruction of the temple on the 14,000th day? Also, that the passages existed centuries before that happened?*"

To begin, we have eight copies of Daniel discovered at the Dead Sea Scrolls site. The oldest copy of Daniel dates to 125 BC. When was the book of Daniel written?

The proposed dates range from 530 BC[5] to 165 BC. Since the temple burned down in AD 70, we know the recorded texts foretell the future well ahead of time. The range is from 215 years to a maximum of 600 years. Either way, these prophetic words are unique. The texts foretell the future.

You may think of asking an important question. "*Are you the only person calculating these dates? Also, how do I know that the time-based prophecy ends with Herod's Temple burning down on the 14,000th day?*"

I am not the only person calculating the dates. Sir Robert Anderson found the precise date for Palm Sunday (April 6, AD 32). He wrote this date into the Christian classic, The Coming Prince,[6] published in 1894. People continue to buy and read that book as it reveals in-depth information about the Bible.

For burning down Herod's Temple, Dr. Jack Finegan[7] calculated the exact date as me (Sunday, August 5, AD 70).

Between the two dates above, there are precisely 14,000 days. The following graphic verifies the 14,000 days, or you can go to that website[8] to check it.

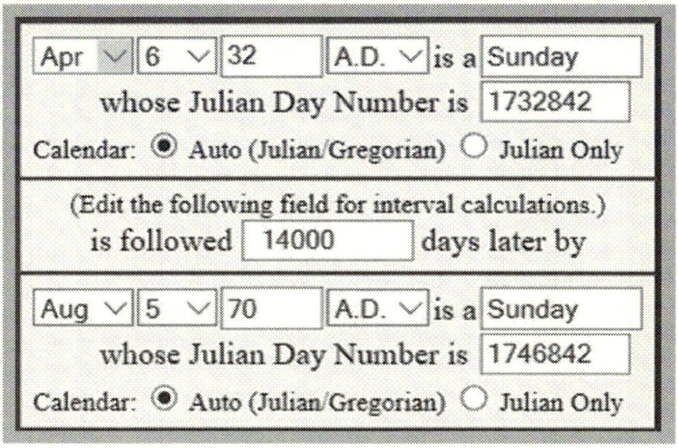

Verifying the 14,000 Days Phenomenon

Importance of the Hidden 14,000 Days

Are you ready to perceive heavenly guidance over human events?

My research suggests divine control of high-level spiritual events. In contrast, there appears to be very little influence on all other events. Because you have the freedom to live as you want.

Regardless, if your soul is at stake, you can expect divine authority in your life. The ideal example is Jesus' death and resurrection, the events that bring you eternal life. Also, you, as an earthling, cannot even begin to imagine how awesome eternal life will be!

Daniel's texts foretell of burning down Herod's Temple. This occurred on the 14,000th day, as the graphic above shows. Finally, you must use the words of Daniel to find exactly 14,000 days based on the foretold events.

There is much more to the time-based prophecy. Because what I am preparing to show you is the 14,000 days sequence requires God-like influence.

Daniel's texts let a highly interested person use a calculator to find the date that Messiah would appear at Jerusalem. You may have known that the Messiah was coming.

In contrast, the prophetic texts do not have numbers to calculate the time from Palm Sunday to the burning down of Herod's Temple. Just to make sure you grasp this, let me write it a different way.

If you lived before AD 32, you may have been able to know the day in the future that the Messiah would enter Jerusalem. Because the time-based texts give you numbers to calculate that day upon which Jesus entered Jerusalem as Messiah.

In contrast, if you lived before AD 70, you would never have known the day in the future that Roman soldiers destroyed the temple. Why?

Because the time-based passages do NOT have any numbers that would permit us to find the date of that future event.

In the graph on the next page, I show the periods where the "time-oriented" texts let you calculate the coming of Jesus. In contrast, you cannot calculate the destruction date for burning down Herod's Temple. The black portion on the right side of the graph shows the unknowable time. That time turns out to be precisely 14,000 days.

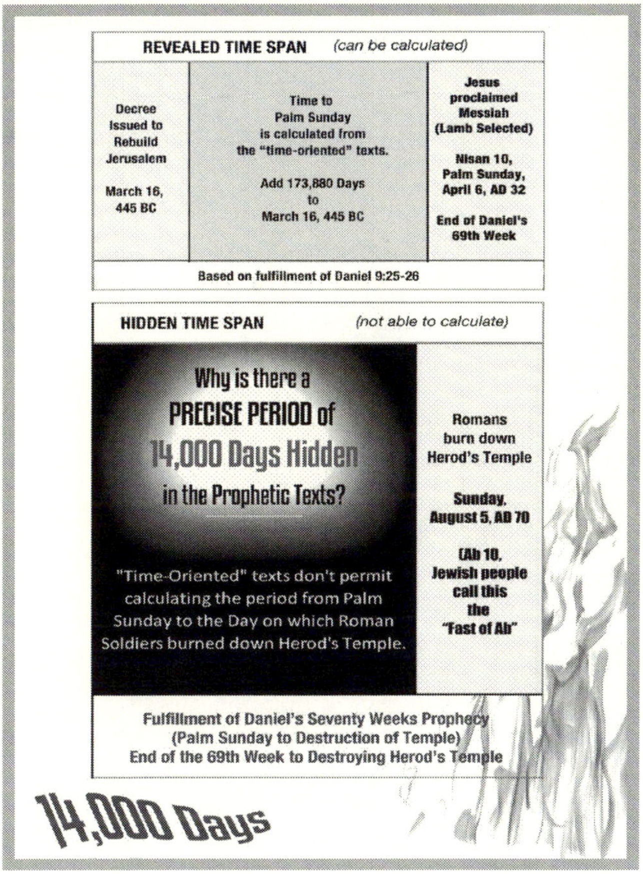

Verifying the Hidden 14,000 Days from the Transcendent Being of the Bible.

~

WE MUST ASK. "*WHY DO WE UNCOVER PRECISELY 14,000 DAYS?*"

The Transcendent Being

The only Being capable of foretelling future events on a precise date is the Creator[9] described in the Bible. Only the Bible has time-based texts for future events.

In contrast, the Qur'an has no prophecy recorded on its pages centuries before events. Both the book of Daniel and the Qur'an record that the angel Gabriel brought them words from God. Since both books claim the same precise angel revealed "holy writ" to each prophet, then the texts should agree!

However, the two books contradict one another. We see and understand the dividing line between the two sets of literature. Because there is no evidence of a Transcendent Source[10] in the Qur'an. You will not find any prophecy in the Qur'an equal to those recorded in the book of Daniel.

This time-based prophecy in Daniel verifies that God exists. The 14,000 days show that God planned the destruction of both Solomon's Temple and Herod's Temple.

What is the purpose of the 14,000-day periods?

To begin with, it shows that God controls events with high-level spiritual meaning. At the center, we find the historical personality, Jesus, the Messiah. His death and rising from the dead are the turning point in history.

Also, the 14,000 days aligned to burning down Herod's Temple reveal a more profound truth. Therefore we can conclude that Jesus' sacrificial death and supernatural resurrection replaced the need for that temple! Said another way, Jesus became the final sacrifice. Destroying that temple appears to be an act of God. On the 14,000th day in the time-based texts, Roman soldiers burned down Herod's Temple. This period of exactly 14,000 days reveals the hand of God at work.

In truth, you can be confident that the death and resurrection of Jesus replaced the need for a temple. The burning down of Herod's Temple validates Jesus' death and resurrection.

It is doubtful that all of this occurred by chance. No human can foretell future events to a precise date that always aligns with 14,000 days. Also, the event on the final 14,000th day unveils high-level spiritual truth. That humans are spiritual at their very core. Because you have a soul that will live forever.

Just as important, the Qur'an appears to be a false spiritual document because its texts teach the diametrically opposite viewpoint. I will begin to unveil Islam in book four of this Expect to Live Forever book series. Look for the title, Secrets - *never heard until now* - of the Spiritual Realm!

Just so you understand, burning down Herod's Temple on the 14,000th day means that Jesus' sacrificial death and supernatural resurrection replaced the temple.

Anointed Brain

FINALLY, WE HAVE THE 14,000 DAYS ALIGNMENT INTO THE TWENTY-FIRST century. This final set comes from scriptures related to the Second Coming of Christ. In fact, the next chapter is about my preaching a sermon on the Second Coming of Christ in 1973. I did not know of the 14,000 days at that time. However, God anointed me in a way that drastically caught my attention. This anointing led to the discovery of the repeating 14,000 days sequences.

As you read into this book, you will find out why I use logic, historical evidence, and the Bible. The *Anointed Brain* experience reported in chapter nine revealed this to me. At the heart of that experience, I found the guidance to use logic, history, and the Bible to unveil the spiritual realm. That chapter reveals why you should read either the second or third book in this Expect to Live Forever book series[11].

What is in it for me?

These studies support that God is extremely interested in you. We, as humans, cannot begin to grasp the Love of God.

He wants you to acknowledge Jesus as the Messiah and as the Lord over your life. That you would open to learning spiritual truth to receive Christ into your life. This means you need to leave your other beliefs behind. Jesus Christ embodies grace and truth. Jesus is the source of eternal life.

This study supports that the time-based biblical texts did, in fact, foretell the future from centuries before to precise dates. No entity except the Creator of the universe could do this. This means that we have "from the horse's mouth evidence" (found in the prophetic texts) that God exists.

You can get inspired when you begin to understand the time-based nature of the prophecy. You arrive at the conclusion that burning down the temple relates to every person that will walk upon the Earth. (*Think about yourself and your friends, then of the burning down of Herod's Temple!*). It appears that God ordained the burning down of Herod's Temple before the creation of the universe. To support this idea, consider a passage from the book of Revelation.

"...written **from the foundation of the world** in the book of life of the Lamb who has been slain" (Rev. 13:8b)

This verse shows that God planned the offering of the Lamb of God, Jesus, even before the creation of the universe. That Jesus' death and rising from the dead link directly to the 14,000 days. We find divine purpose in temple destruction events.

Conclusions

Can we get anything from this quick view of the "time-oriented" prophetic words?

For sure, there is more to reality than our physical beings. You can expect to live forever!

By far, the most exciting aspect of the "time-oriented" prophetic words is the burning down of Herod's Temple. No one could calculate the 14,000 days from Daniel's texts. That is because the passages do not give us the math for that task.

Yet, God planned that period. To achieve precisely 14,000 days required God-like sway over human events. Also, Solomon's Temple burned down on the 14,000th day. These events with perfect 14,000 days aligned show God-like control. These high-level controls only happen for events with high-level spiritual meaning. Are you inspired by a precise period of 14,000 days?

We have learned that the time-based prophecy uses the numbers seven and seventy to foretell the date of Jesus first appearing. In contrast, God hid the 14,000 days until the modern era. This is an eerie sign that God watches you all the time. That God controls events with high-level spiritual meaning. Think of the prophecies in the book of Revelation. There are many high-level events in that book. Please prepare yourself.

In summary, what do the hidden gems of 14,000 days reveal?

1. We know that God exists.
2. That God controls events with high-level spiritual meaning. For sure, burning down both Solomon's Temple and Herod's Temple on the 14,000th day are high-level spiritual events.
3. That God purposely hid the 14,000-day periods until the modern era.
4. It appears that God wants this information spread around the globe before the Second Coming of Christ. Why?

Because God wants you to prepare yourself for an event that will impact everyone upon the Earth. The book of Revelation is mysterious and prophetic. The texts foretell of massive change that will affect every person. Some events will even affect the dead!

Did that makes you gulp hard!

To gain insight into this, I suggest that you set aside time to read and study the book of Revelation. Read that book considering the periods of 14,000 days. Since God destroyed both Solomon's Temple and Herod's Temple on the 14,000th day, what future events will happen soon[12]?

I have chosen to reveal the fact of my out of body experience related to the book of Revelation. I have taken about thirty years to make this decision. Perhaps you think what I have written cannot possibly be true?

The evidence for the repeating 14,000 days is fact-based with support found in two other sources.

1. Sir Robert Anderson's research recorded in The Coming Prince, published in 1894. In recent years, I have found new evidence that validates Sir Robert Anderson's research. I write about this new evidence in the next two books.
2. Dr. Jack Finegan uses data to find the precise date for burning down Herod's Temple. He uses logic, evidence, and astronomy to find the exact date as me.

The data-based approach I use for the "time-oriented" texts makes this information objective for every person that wants to examine it.

I suggest you read everything I write based on the out of body experience and the anointed brain incident described in chapter nine. However, that is up to you. If you begin to freak out, read Appendix A for spiritual guidance and comfort. I will always be praying for you that God blesses you beyond your wildest imaginations.

3

ANOINT!

During the time I was studying to be a pastor at a Christian College, I got involved with church outreach. Leaders placed me in a group of six students, three women, and three men. We would go to local churches in Texas, Oklahoma, or Arkansas. Two of the members were committed to becoming a pastor. The other members were inclined toward music ministry.

During this time, I awoke to the spiritual reality. One moment, I was an average person. The next minute, I became a vessel filled with great-tasting and expensive wine.

Have you ever been in a deep sleep, then suddenly, someone awakens you? In my case, the awakening from sleep affected both my physical body as well as my soul. By this statement, you will learn that there is more to your being than just your flesh. Your body and soul cling together as a single unit.

On this weekend outing, the group traveled to Killeen, Texas. Since I was studying to be a pastor, I preached two sermons on that weekend.

At the Sunday evening service, I chose to preach a sermon on the Second Coming of Christ. I must admit, I did not altogether feel at home with the idea of becoming a pastor on that evening. However, I was about to be spiritually awakened. Little did I know what was in store for me!

I did not know it then, but when I think about it, a student pastor preaching on the return of Christ seems a bit too much. Regardless, it was the specific subject of the Christian faith that enthralled me. It lighted a fascination with me about my faith. I hid a deep interest in the return of Christ. This started because of reading the ten-volume set of Bible stories when I was in the second grade.

On this evening, I did not inform the other five members about the sermon I was going to preach. I was a bit surprised when the ladies in the group sang a song about the return of Christ. The fact that they sang this song without knowing my sermon suggests a great deal about the Holy Spirit. This detail reveals divine leading by the Holy Spirit in and through our lives. Indeed, most people are not aware of God's moving in their life. We are, in general, without a clue when it comes to the spiritual realm. How much are our thoughts and actions at some level under divine guidance?

Are you personally aware of the influence of the Holy Spirit at work in your life?

As the song began, I was sitting on the front pew on the left side of the church. To my right, another student pastor was listening attentively. About the end of the first line of that song, I heard a rustling sound above my head. It was soft but unmistakable for me. Something was going on above my head. Then I heard the word "ANOINT!"

If you were sitting in the pew right next to me, you would not have heard this word. However, I listened to that word as clearly as if you, the reader, were looking straight into my eyes. Then you lifted your right hand as high as possible, paused for one-second, then said loud and very clearly, "ANOINT!" Then you watched as a globe filled with light appeared above my head, then descends into my head and body.

In general these things were elusive to others. It seems this anointing from above was meant only for me. However, some people did see me react to this anointing from above.

As the word "Anoint" ended, I suddenly felt a sphere-shaped ball pulsing with spiritual power beginning to descend on my head. The ball was about eight to ten inches in diameter.

As I began to feel this anointing come down into and through my head, I reacted with a fast reflex. My right hand immediately and quickly shot up to the top of my head. It was a swift reflex action. I wondered, "*What is going on*?"

This spiritual event took me by complete surprise! As my hand lifted extremely fast, my fingers stopped at the top of my head.

I could feel this ball of spiritual anointing coming into my head. I had momentarily forgotten even my surroundings. The student pastor on my right caught the movement of my right hand immediately. Why wouldn't he? His head turned toward me because of my arm movement. The sudden jerk that came out of nowhere, was very noticeable!

To help you grasp this anointing, I have created a graphic. On the illustration, you will see the anointing sphere with my hand positioned near the top of my head.

"Anoint" with Spiritual Ball Descending Phenomenon

I continued to feel the spherical-shaped anointing move down through and around my head. This anointing ball of spiritual power moved at the speed of about four inches per second.

As I had moved my right hand suddenly, the tips of my fingers were now at the top of my hair. I could feel the ball of power had only moved about two inches into my head. I felt its presence deeply as if it were a tangible thing. The anointing continued its path, slowly descending. It stopped when it reached my chest. Then ... it was as if this spiritual sphere opened like a flower in bloom. The ball unfurled, and its presence spread throughout my body. Never in my life had I dreamed or heard of such a thing, let alone experienced it!

The feeling caught me off guard and captivated me. A large tear developed and fell from my face. I was completely aware of what had just happened. I was in awe of the anointing I had just received. I knew at a deeper level that no one other than me in the room had seen with their eyes the change that had come upon me. No one heard the rustling sound or the pronouncement of 'anoint' that I did. No one saw the ball of spiritual power run down through my being. However, to my surprise, I found out later that though the change had not been seen by these other people, they did, in fact, notice me reacting. They knew something spiritual had occurred[1].

After the music, I arose and preached the sermon on the return of Christ. My central scripture came from Luke 21:7, 20-24. Verse 24b is the end of a prophecy by Jesus that the Jewish people would scatter around the world after the destruction of Jerusalem that occurred in AD 70 (in verses twenty to twenty-four). But someday, the Jewish people would return to their homeland. I emphasized that the sign of Christ's return is when the Jewish people retake control of the city of Jerusalem. The twenty-fourth verse takes in the time from AD 70 to June 6-7, 1967. Now it was December 1973, and the clock was ticking.

When I had completed the sermon, the local pastor took over and invited people to come forward for prayer and ministry. About one-third of the attendees responded. I was awestruck!

After the service finished, an elderly man rose to his feet and testified that the Holy Spirit had anointed me. He turned to me, looked me in the eye, and said, *"You were anointed by the Holy Spirit!"* His firm manner told me that he, like me, was utterly convinced of it.

I later discovered that he was the father of Linda, who was a member of my group. She relayed his words to me that her father believed I would have a remarkable change happen in my life.

Since we were in Killeen, Texas, we had to drive back to the Christian College in Oklahoma City. I rode in the back seat and could not sleep. I thought about what had just happened to me. Why did I get anointed?

We arrived back to the campus at 5 am after driving all night. I was so tired that I missed my first class in science at 7:30 am.

Now, think about a student pastor getting anointed with spiritual power. Visualize it happening through a spherical shaped powerful ball, all of eight to ten inches in diameter. Would that get your attention?

Star Trek Correlation

It got my attention, most certainly. I would spend the next fourteen years[2] doing in-depth research driven by the event that I had gone through. Have you ever had your attention captivated by something that caused you to change course in your life?

If you like the old Star Trek shows, I recommend an episode with this theme. Please watch "Where No Man Has Gone Before." This is season one, episode four. Sometimes you can watch it on Netflix for free.

This Star Trek episode also relates to chapter six. When you read about the experience in chapter six, you can begin to understand why I have become hooked on the study of ultimate reality. Just so you know, "Ultimate reality" entails what will happen to you when you die.

Conclusions

Can we draw any conclusions from this anointing experience?

For sure, you would have seen me react with my right hand to the spiritual anointing that came down on my head. You would have noticed the tear that came out of my eye. Perhaps of greater importance, you would have responded to the Holy Spirit as you saw other people react to the logic and evidence used in that sermon. You can get all the evidence, the logic, and details in either the second or third book of this Expect to Live Forever book series.

Finally, this story from the heavenly realm may open your heart and mind to the idea that you can personally know God. That in truth, you can have a wonderfully deep relationship with God. Think about Jesus' words.

> "Ask, and it shall be given to you; seek, and you shall find; knock, and it shall be opened to you" (Matt. 7:7).

4

MY CALL TO WRITE

I SPENT YEARS IN DEEP THOUGHT ABOUT THE ANOINTING AND THE sermon I had preached in 1973. What did that sermon genuinely mean for me? Perhaps for the entire world?

Please recall, that sermon included texts that foretell the future. For me, it was essential to find out how well those texts foretold the future. I came up with the idea that I may find an answer by getting the exact dates of events predicted by prophetic words. I knew there is only one "time-oriented" prophecy in the Bible. I wondered if the foretold events were accurate? Would I find a pattern that repeats?

From the beginning, I had ideas that turned out to be correct. It seems that the anointing I received, as revealed in chapter three, somehow affected how I think!

I began with the idea of finding the length of a biblical generation. I had decided that the time of thirty-eight years and four months could be the answer. This idea is what led to the discovery of 14,000 days hidden in biblical texts. Two sets of scripture give support to this idea.

The first idea came from passages related to Moses. Those texts were about an entire generation of people dying in the desert. The period for this to occur was thirty-eight years. Consider this verse.

> " . . . the time that it took for us to come from Kadesh-barnea, until we crossed over the brook Zered, was **thirty-eight years**; until all the generation of the men of war perished from within the camp, as the Lord had sworn to them" (Deut. 2:14).

The second set of thirty-eight years came from early Christian history. I came around to believing that Jesus died on the cross in AD 32. If Jesus died on the Passover in AD 32 and Roman armies burned down Herod's Temple in AD 70, this is another set of thirty-eight years and some months.

At that time, my intuition linked those earlier generations at the time of Moses and Jesus to the modern era. For example, the thirty-eight years at the time that Rome burned down Herod's Temple. Would I find the same period for the return of the Jewish people to their homeland in the twentieth century?

The fact of the Jewish people returning to their Promised Land was based on the sermon I preached in 1973. Also, there may be something unique about the prophecy of Jesus fulfilled in the Six-Day War. I was bearing in mind that I would find a repeating time based on that sermon. Please bear in mind that the anointing I felt back then, now glued the biblical texts to the facts I spoke of in that sermon. This clue kept me digging for the answer.

The idea that Jesus died on the cross in AD 32[1] came from The Coming Prince. Sir Robert Anderson published it in 1894. In this book, Mr. Anderson had concluded that Daniel's "time-oriented" prophecy pointed to Palm Sunday, April 6, AD 32. Sir Robert Anderson had derived this date by use of the positions of the sun, the moon, and the earth in ancient history. Then he used the calculations found in Daniel's "time-oriented" prophecy. The date of April 6, AD 32, was an easy date for me to remember because my mother was born on April 6, 1925.

The "time-oriented" words also foretell of the temple's ruin that happened in AD 70[2].

To find the date of the loss of Herod's Temple, I used the eyewitness account of Josephus. He wrote of the temple's ruin as a mystical event[3]. You may find why Josephus wrote about the temple in a mystical way a bit intriguing. Every time a foreign army burned down the temple, it happened on the exact same day on the Hebrew calendar. To Josephus, he believed this was an act of God.

I, too, have come to the same conclusion after decades of research. We will find a good number of mystical stimuli to support God plans and controls temple destruction events, because those events have "high-level spiritual meaning."

Back to finding the date that Romans burned down Herod's Temple. Can we know the exact date of an event in ancient history?

After years of research, I have concluded that if we can find the day of the week on which an event in ancient history occurred, then we can get the exact date of a specific incident. Since Josephus was an eyewitness to the destruction of the temple, he gives us a clue as to the day of the week that Romans burned down the temple. Josephus writes that the Jewish people rested on the day before the temple's destruction. This would mean that the day before the temple's removal was a Saturday, the Sabbath. Therefore, the next day is Sunday, the first day of the week. This information from Josephus strongly suggests that the temple burned down on a Sunday.

Josephus gives us a second clue. He writes that the temple burned down on the tenth day of the fifth Hebrew calendar month. This means that ten days had passed since a new moon. Therefore, we should find ten days from the new moon line up with a Sunday.

To my amazement, I discovered that the ten days since a new moon aligned perfectly to a Sunday. The precise date using the day of the week comes out to be Sunday, August 5, AD 70. Now, this is a date that is easy to remember for me as well. Because of the fact, my father was born on August 5, 1917.

I have included the following graphic of my parents' gravestone. I did this so you can verify their birthdates are the same as I have written into the current chapter.

Gravestone with Birth Dates Phenomenon

I NEED TO POINT OUT THAT I AM NOT THE ONLY PERSON THAT CAME UP with these dates. I have referenced Sir Robert Anderson as the source of the April 6 date. For the destruction of the temple, Dr. Jack Finegan[4] calculated the August 5 date. Also, we have three people[5] from ancient history that verify Romans burned down Herod's Temple on a Sunday.

I did this research in the 1980s. Why would my parents' birthdays be the same dates as found in the only "time-oriented" prophecy in the entire Bible?

I found it funny, perhaps even mystical. Does it mean that I was prepared to do this biblical study, even before I was born?

Even deeper, it appears the Holy Spirit destined my parents to know one another and to raise their family? I am sure that they did not have a clue about the Holy Spirit directing their lives in this way. Or is this all simply a coincidence?

Maybe you are into astrology, which means the day of your birth influences your life. Do you think the things I have just written support that viewpoint.

Biblical texts confront that viewpoint. The biblical prophet Isaiah gives the biblical view of these ideas from about twenty-seven centuries ago as follows.

> "[13]Let now the astrologers, those who prophesy by the stars, those who predict by the new moons, stand up and save you from what shall come upon you. [14]Behold, they have become like stubble, fire burns them; they cannot deliver themselves from the power of the flame; there will be no coal to warm by, nor a fire to sit before" (Isa. 47:13-14)!

Just so you understand my position, there is much more to my research than my parent's birthdates. In the early years of my study, this information aided me. Perhaps it helped me to continue with the investigation, because both my parents' birthdates aligned with the dating of ancient events in Daniel's "time-oriented" prophecy.

For me, I view the mystical alignment of relevant birthdates from a biblical viewpoint. That is to say that the Holy Spirit is much more involved in our lives than we can begin to realize. I must insist that the aligning of birthdates have nothing to do with astrology.

Maybe you are intrigued by the information above. After you have read this book, you can arrive at your own conclusions. Do you have insight into the Holy Spirit moving in your life? For example, why are you reading this book?

On the first Sunday evening of 1987, I was at my parents' home from Christmas to New Year's Day. It was my Christmas vacation. I had already assembled the dating, as noted above, which is still valid today. I was not willing to share this information with anyone. I kept it completely secret, locked up in my paperwork, and in my mind. However, the Holy Spirit would intervene in my life again. More than thirteen years had passed since I had heard the rustling above my head with the word, "ANOINT!" Then the sudden descent of a sphere-shaped ball of spiritual power to anoint me from above.

I went to church on that Sunday evening. My reason for going to church was to see people I had known for years. I now lived over 500 miles away. The church attendance had gone down as there were only about twenty-five people in that service.

At the church service, everything was normal until the local pastor called me out in the back pew. He asked me to give my testimony.

I stood up, not knowing what to say. I said that it was nice to be home for the Christmas season and to see everyone again. Then I said something that I had not planned to say.

"I have been thinking about writing a book about the Bible. Will you pray for me for guidance?"

I sat down and said to myself, "*You idiot, why did you say that*?"

There was a second testimony, and then Pastor Nelson asked us to stand for prayer. I stood in the back pew, twiddling my thumbs and waiting for the prayer to end. When Pastor Nelson began to pray for guidance for me to write a book on the Bible, the Holy Spirit manifested once again. At the instant he began to pray; I felt a three-inch band of spiritual heat come across my chest from one side of my lowest ribs to the other side. The length of the fiery breastplate was about fifteen inches. The breastplate was just above my liver. I felt this breastplate of fire for nearly an entire minute as Pastor Nelson continued to pray for me to have guidance in writing a book. At the instant that Pastor Nelson stopped praying for me, the breastplate of fire stopped.

The graphic below illustrates what I felt. A breastplate made of fire across my lower chest, guiding me to write the research that is found in the second and third books in this Expect to Live Forever book series.

Holy Spirit Guidance Phenomenon

I TOOK IT AS A SIGN AND FIRMLY SAID TO MYSELF, "*I MUST WRITE A BOOK about my research!*"

Conclusions

CAN WE DRAW ANY CONCLUSIONS FROM THIS EXPERIENCE OF MY CALLING to write about the spiritual realm?

For sure, you would have seen me place my hand palm up close to my chest. However, you would not have known what I was experiencing. I told no one about this event for several years as I kept the details to myself.

You can review the gravestone birthdates and compare them with the precise dates for the time-based prophecy. You will find that they agree to the exact day. This begins the objective evidence that points to the spiritual reality.

5

"I WANT TO SEE THE POWER OF GOD"

How close can you get to God?

I doubt anyone knows the answer. To begin with, most people have not a clue. What if you continue to grow in faith as newer and bigger experiences keep coming your way. When you find that you have a continual relationship with God and feel, "the presence" all the time, you can begin to answer the question above.

However, even that does not reveal the answer. Perhaps you need to realize, God is infinite.

In this chapter, I disclose a life-changing incident that took me to a level I had never dreamed of. Are you ready for an eye-opening event?

In January 1990, I went to a fellow believer's home. We were starting a Bible study for singles. At that time, I was attending a Presbyterian Evangelical Church, which had about 5,000 members with about 500 singles. Single people would come to the church to meet other singles. This church was a comfortable and thriving place for singles to meet.

Since it was a large singles group, I did not know the co-leader that well. For the first meeting, we met to discuss what part of the Bible to begin the Bible study with. Also, we wanted to learn more about each other's faith.

Before I continue, let me ask you a question. Is there an on/off switch that instead of controlling a light bulb, switches on your soul to awaken you to the supernatural realm?

Flipping the Spiritual Switch from Off to On

I REVEALED TO MY CO-LEADER WHAT HAD HAPPENED TO ME IN CHAPTERS three and four, to which she said to me. *"If you lay your hands on me and pray for me, I will be healed!"* She asked with faith that the light switch was going to switch from off to on, then bring the desired healing.

"I don't know," I responded. *"I have never done this before!"*

"Well, you don't have to do it." She said quietly as if she felt let down!

However, as the first word from her last sentence began to flow from her lips, sparks of hot fire started to go on and off in both palms of my hands. By the time she finished the sentence, at least twenty or more sparks of fire had flashed in each hand. The flashes of fiery sparkles turned on, then off in both palms of my hands. The graphic below illustrates these twinkling hot sparks inside my palms! I had never experienced or even heard of it before this happened to me.

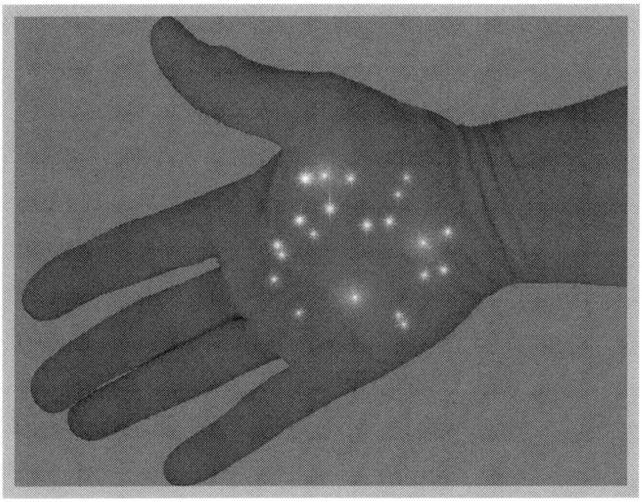

Right Hand with Twinkling Hot Spiritual Sparks Phenomenon

IMAGINE THAT TIME STOPPED FOR A FEW SECONDS JUST AS SHE FINISHED the sentence, *"Well, you don't have to do it!"* You would find yourself being the central actor in a Twilight Zone movie. Instead of watching the big screen from a theater seat, you are one of the characters on the screen. The switch controlling the spiritual realm suddenly turned to the on position with twinkling hot sparks in your palms.

It is sort of like standing in a room that is pitch black. Then the door begins to open, and through the widening crack in the doorway, intensely bright rays of light begin to shine through. You lift your arms to cover your eyes from the intense spiritual light. Will you step through the doorway into the brilliant, radiant presence? Are you ready to leave the everyday mundane existence of a human being?

I looked into her eyes and said factually and bluntly, *"Ask, and you will receive!"*

"How do you want to do this?" she asked.

"Get a chair and place it in the middle of the room," I instructed her.

From the time she had said, *"Well, you don't have to do it,"* both my hands had continued to have twinkling hot sparks turning on, then off and then on again. Also, I felt the fiery spiritual sparks at different locations in the palms of my hands.

Each spark would start as a tiny dot. Then grow to about an eighth of an inch in diameter, sort of like the shape of circular flashes of light. Each spark would last only about three to four-seconds. Then each spark would fade away as others started in different locations. This sensation was entirely beyond me as I was experiencing the presence of the Holy Spirit in an unusual way. For sure, I was captivated.

She sat down on the chair, and I gingerly placed both of my hands on her head. *"There is definitely no need to push because this is not of me,"* I mused in my mind. In regard to applying pressure, I only put the weight of my hands on her head. For sure, I used no physical force.

When I placed my hands very gently on her head with the twinkling hot sparks continuing to fire on and off inside my palms, a thought went through my mind. I said to myself just before praying aloud for healing, *"I want to see the power of God!"*

Then I said a brief prayer. I would say that the length of the prayer was less than fifty words. The time length was about a minute at the very most. The response was immediate.

What happened to each of us was mind-blowing! To say that it was well beyond major surprise would be an understatement! It would seem like I had stepped out of the real world into a fairy-tale. However, that fairy-tale world is the reality. The spiritual realm is the world of miracles and dreams come true. Can you begin to conceive that reality[1]?

As the very first word of my prayer began, I became captivated! The instant I uttered the first word, my life changed. I felt the equivalent of hot liquid oil flowing out of my wrists and hands into her head.

This intensely warm healing oil flowed from my wrists, down

through my hands, then out of my palms. This healing oil felt like a column of hot spiritual oil, leaving my hands. The temperature was about 120°F. I watched my hands turn red from the soothing heat. Also her head turned a red hue as the anointing flowed into her head.

I could also feel the hot oil going down into her head. It was as if I had nerves inside the holy oil and felt where it was going.

The graphic below is an attempt to illustrate what happened. If your head were below my palms, you would have felt this column of hot oil flowing down into your head. Also, I sensed this hot liquid oil inside her head.

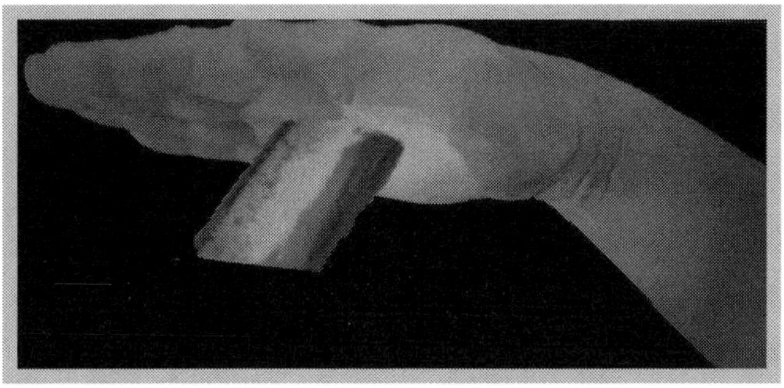

Healing with Hot Liquid Oil Phenomenon

PLEASE UNDERSTAND THAT THERE WAS NO NATURAL OIL. INSTEAD, IT felt like soothing hot oil. In truth, the entire feeling was utterly spiritual because we are spiritual beings made in the image of God.

For two people to experience the same spiritual event as two believers in Christ is genuinely remarkable. We would each describe the incident in the same way with a slightly different perspective. I have described what I felt in the paragraphs above.

After lifting my hands from her head, I pulled up a chair and sat facing her. She said to me, "*Your hands became really hot!*" She had a charming smile on her face as she looked into my eyes.

My recall became uncertain due to the power and overwhelming evidence of this spiritually driven event. At the right time, I asked, "*Why do you need healing*?"

She answered that she was having heart pain. It was from the fact that she was divorced and she had emotional agony from her loss. To make a long story short, healing took place.

When I drove home from that spiritually driven event that evening, I was excited and captivated. It was awesome to have such a dynamic spiritual experience. It was after nine pm when I arrived home. I at once went to bed and fell asleep. In the morning, I was back to my every day mundane life. My hands were not burning with the fiery presence of the Holy Spirit anymore.

At that time, I was a technical writer for an aerospace firm. My boss was a person that had studied to be a pastor but had many personal problems. So, he never became a pastor. Regardless of his problems, I told him about what had happened. He responded that he used to go to a Charismatic fellowship and had experienced similar things in his life.

It was attention-grabbing to me since I had not grown up in that type of church. I had just begun to open my mind and heart because of a remarkable spiritual experience. Imagine a hurting person getting healed!

My day at work was typical. I wrote technical documents and then drove home after the shift ended. My typical day to day routine was to go to my residence, then lay down for a thirty-minute nap. On this day, I did not expect to get another surprise from the spiritual realm. When I opened the door into that apartment, I did not know that my life would be changed completely.

When I lay down to take a nap, I usually fall asleep in less than five minutes. I rest deeply for a short period, and then feel ready to get up and go for hours.

On this day, I went to sleep without any fiery spiritual presence in my physical body. However, when I awoke, my hands were on fire; they were red-hot, to the point of almost hurting. It was noticeable and I was awestruck.

As it turns out, what happened to me on that evening, turned into a life-changing event! Ever since that day, I have always felt the fiery and tingling presence of the Holy Spirit. I live my life in the presence of God. Because the spiritual part of us (our soul) is intimate with the fleshly part of our being[2]. The Eternal One created us in the image of God according to scripture. In the New Testament, we find a verse about Christ's followers as being three in one, just like the Trinity.

> "...may the God of peace Himself sanctify you entirely; and may your **spirit and soul and body** be preserved complete, without blame at the coming of our Lord Jesus Christ" (1st Thess. 5:23).

A person that does not believe in Jesus only has a body and a soul. That is my personal perception based on my encounters and what I read in biblical texts. To help clarify this idea, when you believe and accept Jesus, you are born of the Spirit, which creates your spirit. You change from a body and a soul; to a body, soul, and spirit.

How can you understand the Trinity?

A verse in Matthew reveals one God in Three distinct beings. Each Being is somehow unique and independent. However, they are entirely intimate as a monotheistic Being. You may find it interesting that tests in quantum physics[3] have uncovered concepts that help us gain insight into the Trinity.

"[19]Go therefore and make disciples of all the nations, baptizing them in the **name** (*singular*) of **the Father and the Son and the Holy Spirit** (*three unique beings that have a single name as One God*), [20]teaching them to observe all that I commanded you; and lo, I am with you always, even to the end of the age" (Matt. 28:19-20).

Now back to my personal experience. I always feel the presence of the Holy Spirit in my life. Even if I am far from perfect and do sinful things, the Holy Spirit abides in my life. This is the grace and mercy of God as He walks with us through the valleys, the hills, and the mountain tops of everyday life.

Since I am not perfect, like you, I often recall the following verse.

"... if we walk in the light as He Himself is in the light, we have fellowship with one another, and the blood of Jesus, His Son cleanses us from all sin" (1st John 1:7).

The experience described in this chapter was one of the most life-changing ones in my entire life. I think about it often. In fact, when my father passed away, I spoke about this experience at his funeral. I remember calling my father and telling him about this event in my life. When my father heard the story, he said to me, *"Oh, you have the touch!"*

That response from my father reveals a great deal that I did not know about him. He must have experienced the presence of the Holy Spirit in genuine ways in his life, as well. For sure, the spiritual event described in this chapter was another "taken by complete surprise" type of event. Because these were not regular occurrences by any stretch of the imagination for me.

Conclusions

What can we conclude from this healing event?

For sure, we know that there is a spiritual realm. This experience teaches that each of us has an eternal soul. Also, we are to take care of one another and to help one another.

Finally, the Bible is the Word of God. This is my conclusion because we met to discuss the Bible. That discussion about the Bible led to her healing.

If you were present in the room, you would have witnessed several items. You would have listened to our conversation. When I prayed, you would have seen both my hands as well as her head turn a red hue from the spiritual healing power coming out of my wrists. If you shook my hand, you would have felt the temperature was extremely feverish. A non-believer would ask, "*Are you okay*?"

You would have seen events related to prayer in the name of Jesus.

Also, you could ask questions of two people that experienced the reality of the spiritual realm. Then you can ask yourself. "*Am I ready to learn the fact that ultimate reality is spiritual*?"

6

PROPHETIC CALLING

IN CHAPTER ONE, I WROTE ABOUT GOD CALLING ME TO BE "*ONE OF THE* ***prophets of the book of Revelation!***" I supplied some details of this calling. The validation included my spiritual arms and hands rising out of my physical body with a voice of praise.

Indeed, an out of body experience is usually related to New Age people. However, New Testament verses support that God spoke to the Apostle Paul through an out of body experience. We find in 2nd Corinthians 12 the supporting texts. The least of the Apostles, Paul, received divine revelation when God took him to paradise.

> "1... I will go on to visions and revelations from the Lord. 2I know a man in Christ who fourteen years ago - whether in the body I do not know, or out of the body I do not know, God knows - such a man was caught up to the third heaven. 3And I know how such a man - whether in the body or apart from the body I do not know, God knows - 4was caught up into Paradise, and heard inexpressible words, which a man is not permitted to

> speak. [5]On behalf of such a man will I boast, except in regards to my weaknesses" (2nd Cor. 12:2-5).

While studying the entire text of 2nd Corinthians 12, I find it most satisfying to read the following passage. The Apostle Paul wrote.

> "...everything we do, dear friends, is for your strengthening" (2nd Cor. 12:19 NIV).

I hope my writing in this book series strengthens your faith too. That you understand the Bible alone has words from God. He knows all the private details about every person on planet Earth. That each of us is PRICELESS in God's view as recorded in the New Testament, which is the final revelation of God.

Also, you understand that the out of body experience is not about me.

Do you go to a church where the believers like to sing and raise their hands and arms in praise as a form of worship? If so, then you may gain a better understanding of the out of body experience.

I awoke at approximately 2:30 am on February 1, 1990. Only sixteen days had passed since the healing experience that I wrote about in chapter five. As I opened my eyes, I felt that my hands were on fire. They were churning with the presence of the Holy Spirit evident. As a side note, I had awoken the night before at about the same time. On both nights when I had woken up, my hands were on fire with the presence of the Holy Spirit. On this night, it was going to be different from the previous night. I never knew when an enjoyable experience would occur.

On this night, I fell asleep with my stomach on the couch. I had my arms and hands positioned above my head on the living room couch. That is the way I used to sleep.

When I awoke, I lifted my head up off the couch. Then I dug my elbows into the sofa enough to lift my upper chest slightly above the lounge. My hands pointed toward each other. The fiery presence of the Holy Spirit moving in my hands caught my attention. It was as if I wanted to look at this Holy Presence in my hands.

Then the thought went through my head, ***"You are one of the prophets of the book of Revelation!"*** At this time, I must point out that the sentence sounded like I was talking to myself. The thought pattern was my own voice.

At the end of this sentence, everything changed! I received pure knowledge about the spiritual realm. I instantly knew that every human has a soul and that we can expect to live forever! This new experience of the spiritual realm took in only three-seconds. However, it was more than enough to awaken me to the absolute certainty of the spiritual realm.

At the end of the phrase, "... of the book of Revelation," two things happened. My spiritual hands and arms rose up out of my physical hands and arms. At the same time, a voice of praise filled my mind. As my spiritual hands and arms moved to where those hands pointed straight up toward the heavens, the voice of praise accompanied their movement. The voice of praise was the word, "WHEW!" This praise word was an extended word that lasted about two and a half seconds.

To help you understand this spiritual experience, I have chosen to divide the movement of the spiritual appendages from the voice I listened to with my mind.

Movement of Spiritual Appendages

At the end of the phrase, "... of the book of Revelation," my spiritual appendages moved rapidly to a position above my physical head. The time that my spiritual hands and arms lifted to their maximum height was about three-tenths of a second. As they reached the maximum height above my head, my spiritual arms and hands stopped moving. They remained in that position with both of my spiritual hands pointing toward the heavens. My spiritual hand and arms stayed in that position for two-seconds. During that time, the voice of praise was ongoing and came to an end.

At the end of the praise, both of my spiritual hands and arms moved down into my physical hands and arms. The entire time that the spiritual arms and hands were above my physical appendages was about three whole seconds.

The Voice of Praise

At the instant that the spiritual appendages began to lift out of my physical arms and hands, the voice of praise began. I listened to the praise with my mind.

I have already mentioned that the thought, *"You are one of the prophets of the book of Revelation"* was my voice. That sentence sounded like I was talking to myself.

In contrast, when my spiritual appendages began to move, the voice changed. As the praise word, "WHEW" began, the voice changed to a voice I have never heard my entire life. No human has ever spoken like this to me. This voice was not from an earthly source, but rather, from a heavenly source. I cannot even repeat it. The intonation that spoke this word was from another world, like the wind blowing. Maybe it is best to say it was the voice of a Holy Wind, a divine voice.

I listened to the praise word, "WHEW!" that lasted about two and a half seconds. Perhaps it is worth noting that this voice of praise only occurred during the time my spiritual appendages were raised above my physical appendages.

My Physical Appendages

I MUST EMPHASIZE THAT MY PHYSICAL APPENDAGES DID NOT MOVE during the out of body experience.

I created a graphic to help you grasp what happened to me. In the illustration below, you can see my physical hands pointing toward each other. I must stress that my physical hands and arms did not move during this entire experience. Also, that my physical hands stayed in one position, pointing toward each other. I did not have time to think about moving my physical appendages. My mind became absorbed in the out of body experience with the voice of praise!

I did not initiate this praise with uplifted spiritual hands and arms. In my view, this happened in a way that I would know the intent of the revelation. That God set me apart before I was born to be one of the prophets of the book of Revelation.

In the illustration below, you can see the grayscale arms and hands that represent the spiritual parts. Please notice the way my spiritual hands changed their direction to point toward the heavens. During this praise be to God position, I listened to the voice of praise with my mind.

Spiritual Arms and Hands Contrasted with Fleshly Appendages

I THINK OF MYSELF AS A NORMAL EVERYDAY HUMAN BEING. I AM NOT A person that likes to go to a church service so I can shout out "Hallelujah!" or "Amen!" to things said or other acts of humans during a church service. I am not such a person. Instead, I grew up reading. If I had my way, I would just live at a library. I like to investigate or research things and use mathematics to derive and reach credible conclusions.

In 1973, the Holy Spirit anointed me just before preaching a sermon on the Second Coming of Christ. I spent a lot of time in deep thought. I wondered "why" the Holy Spirit anointed me in 1973? That is why I spent a lot of time in libraries doing research that it seems no other person ever did before me.

In my view, it is vital to understand this out of body experience about the book of Revelation. What does this reveal about ultimate reality?

Feeling your spiritual arms and hands reveals that ultimate reality is spiritual. You can expect to live forever!

When your physical body dies, your soul will live on. I have first-hand knowledge about the spiritual realm based on this experience.

The sentence about the book of Revelation[1] means that only the Bible has the word of God.

Debate with My Atheist Friend

YOU MAY ASK, "*HOW DO I KNOW I HAVE A SOUL THAT WILL EXIST forever?*"

I look you in the eye and say. "*When my spiritual arms and hands lifted out of my physical parts, without question, that verified the spiritual realm for me. This means that you also have a soul that will live forever!*"

After a short pause, you stare at me to ask, "*That was you. In contrast with you, I did not have that out of body experience. How would I know you are telling the truth?*"

Instantly, I say with confidence. "*There are many ways of validating my experiences. One of those would be where other people saw the same event with their eyes. Also, they heard that event with their ears. This is an incident I like to call the Psychic Bird Story.*

Five people became eyewitnesses of a spiritual event. The observations of those people allow us to calculate the reality of the spiritual realm with a high confidence level. There is only one chance in 10,000 that the bird deceived us. You can read about the Psychic Bird story in book four of the Expect to Live Forever book series.

"*Oh, wow! That is a huge number to back up your worldview. However, I am an Atheist. I cannot even think about there being a spiritual realm. Evolution explains everything to me!*"

I pause for only a second to make sure you finished speaking. "*Since you are an Atheist, you need to read either book two or three in the series. Because the author uses known science to reveal that Atheism leads to nowhere. In fact, known science points Atheism back to the supernatural realm!*"

I watch you go into deep thought. "*Hmmm*!" is all I hear as I continue to watch you in deep thought. Your eyes squint as you continue to think. "*This guy talks about an out of body experience related to the book of Revelation! Also, he has confidence in it. I wonder if this can be the reality*?"

I continue. "*How much faith does it take to believe what Atheism teaches*?"

"*I am beginning to get your point. Atheism bases its ideas on a natural universe that must be eternal! Even if there are many natural universes, each of them will only run out of energy and die. I wonder how this all got started*?"

Instantly I respond. "*For sure, Atheism requires you to believe in magic! You must assume there is no such thing as the supernatural realm that will never end. In contrast, the Atheist worldview requires a natural universe to be eternal. However, known science shows that natural things do not last forever. Our universe will stop working someday as the stars use up their atomic energy*!"

"*Ohhh, You're right. I have been thinking about that. I must put a lot of faith in what Atheists teach since it has problems. I grew up in a Christian home. My worldview changed when I went to college. I think I want to know more about these 14,000 days sequences that go from the time of Moses into the twenty-first century. Maybe there is more to this than I can fathom at this time*!"

I smile as I say, "*The evidence of the 14,000 days is why I had an out of body experience about the book of Revelation. You are beginning to understand something clearly at last. These ideas are based on logic, historical evidence, and biblical texts. Please study these ideas and take the time to look into them*!"

"*What would it mean if I didn't take time to investigate*?"

"*That would mean you have chosen to live by a paradigm that may, in fact, be misleading you. It would be the same as letting your worldview hold your mind captive. Are you willing to think things through? A great example is a paradigm held by scientists into the twentieth century that began thousands of years ago. They believed and taught that the natural universe was eternal. That paradigm took thousands of years to oust. If you have a closed mind, then you are open to false paradigms.*"

The logic I spoke begins to open your mind!

You respond, "*Great answer! I am going to read all your books. Your viewpoint is unique and fascinating! Imagine that, a guy with an out of body experience about the book of Revelation in the twentieth century that uses science to explain reality! It would be very cool, indeed, to live forever*!"

Let's Analyze the Disclosure

"You are one of the prophets of the book of Revelation!"

In this section, I will dissect the meaning of this sentence, bit by bit. I must emphasize that the out of body experience verifies the spiritual realm.

The sentence begins as: "*You are one of ....*"

This reveals that I received a revelation from the Holy Spirit. This message did not come to me from me. If the thought pattern had come from me, the sentence would have begun in this way. "*I am one of ....*" Also, there is the fact that there would not have been an out of body experience. Only the Holy Spirit would cause the out of body experience.

Now consider the phrase: "*...one of the prophets...*"

Notice that the phrase suggests that I am only "*one*" of many other prophets. This phrase means there are others called to the same position in the global church and that the task is not just mine.

Now consider the phrase: "*...of the book of Revelation*."

I must emphasize the use of the word "OF" instead of the word "IN." That means I am NOT a prophet "IN" the book of Revelation. I am not a person foretold as a character in the book of Revelation. Is there a logical answer to what this means?

Logic suggests that I am a prophet that precedes the book of Revelation. Maybe I am to warn of the carrying out of the book of Revelation. This makes sense to me because of the many spiritual experiences that I have had since 1990. I have lost count of the vast number of these encounters. However, these experiences help to explain the visions in the book of Revelation. I must emphasize that I am NOT a prophet IN the book of Revelation.

Just so you know, let me insist. I am not a person expected by those prophecies. I am *not* one of the 144,000 people labeled in Revelation, chapters seven and fourteen. Also, I am not one of the two witnesses described in Revelation, chapter twelve. I do not have spiritual powers related to such persons.

Again, let us reflect on the phrase "*...one of the prophets...*"

These words would strongly suggest that God called me to be a prophet. What is a prophet?

There seems to be a great deal of variation about what a prophet does. The most basic idea is that God reveals spiritual truth through a true prophet. In contrast, there are also false prophets. These are the people who claim to hear from God but are religious deceivers. It is okay if you ask of me, "*Are you spiritually deceived*?"

I would answer no, based on the further dissection of the entire sentence. Also, based on what I wrote in chapter two as well as in either books two or three. Please read or listen to the audio for Eternal Life - *Why you should expect to live forever*! Or you may choose to read the more in-depth version. Beyond - *The Coming Prince by Sir Robert Anderson*.

Let us now consider another way of looking at the sentence. "***You are one of the prophets of the book of Revelation.***"

The word "*Revelation*" in the original Greek language is the word "*Apocalypse.*" The meaning of apocalypse is to "*unveil.*" Therefore, let us change the sentence to read, "***You are one of the prophets of the book of the Unveiling.***"

This is the best way of gaining insight into this spiritual experience. What it means for you and me? Will it help us understand?

Please consider that chapter four is about God calling me to write. When that happened, I only wanted to write of the 14,000 days hidden in the texts of Daniel. There are two crucial facts we learn from this research.

First, the study shows the historical events that occurred. These events unveil the hidden 14,000 days.

The second fact is that Roman soldiers burned down Herod's Temple on the $14{,}000^{th}$ day. In chapter two, we also learned that no one could calculate from Palm Sunday to the day Romans burned down Herod's Temple. The 14,000 days hidden in the texts reveal God-like control of human events. Like saying. "*God controlled humans so that Romans soldiers burned down the temple on the $14{,}000^{th}$ day*!" Why the $14{,}000^{th}$ day?

The answer is that Jesus' death and rising from the dead made the temple obsolete. Because Jesus, the Messiah, is the final and ultimate sacrifice. Jesus gives us eternal life. There is no need for animal sacrifices, which is more evidence of the spiritual realm. It is a message of hope and the reality of eternal life.

Add to that the out of body experience of the book of Revelation. When you die, your soul will also rise out of your body. You will know that. You will feel your soul just like you feel your physical body. We know that ultimate-reality is spiritual.

Let me ask. "*How do an out of body experience and the study of time go together*?"

To begin, God exists beyond time. God transcends time. God knows all future events. The out of body experience about the book of Revelation is related to foretelling the future. These two ideas go together like clay pressed into a mold. God knows the future and the book of Revelation foretells the future. Think about it.

As a result, we "unveil" that the Bible is the Word of God. People who continue to disagree base that on their paradigm! If that is you, consider the question. "*Am I spiritually deceived*?"

My research to unveil the 14,000 days is the result of a detective story. I found evidence that reveals God guides human events. However, we find limits to this high-level guidance. Because it only occurs when the final incident on the 14,000th day reveals spiritual truth. We learn about finely tuned spiritual truth from these events.

How do I view myself as a prophet?

God called me to unveil that God exists! That God is in control! That we have evidence of the spiritual realm unveiled. That the Apocalypse will change the entire planet.

Finally, consider that before times of great trouble begin, God always sends a prophet. That person forewarns of coming disaster. I write this because I had an out of body experience with the statement. "***You are one of the prophets of the book of Revelation.***"

> "Surely the Lord God does nothing unless He reveals His secret counsel to His servants the prophets" (Amos 3:7)[2].

What does my "out of body experience" about the book of Revelation mean based on the text in Amos?

I have reported it. It is up to you to prepare yourself. You have been issued a warning about the future. I strongly suggest that you take the time to read and study the book of Revelation as well as the Bible.

Prepare yourself by waking yourself up to the "fact" of the spiritual realm. For sure, you will live forever! For sure, you must prepare yourself. To prepare yourself, I have included Appendix A so you can gain insight into ultimate reality.

You can be ready for any and all things that the future will bring. Take courage that what you need to do is extremely simple. Because a Supernatural Being does not want you to fail or that you must jump through hopeless hoops.

In contrast, all other religions require you to become "good enough" to enter eternal life. You must jump through the hoops correctly and with great perfection. The biblical worldview exposes such beliefs as false.

What must I, the author, do?

I must share the message of the words as I received them. That I must write about the fact of the unveiled 14,000 days based on logic. In support of this viewpoint, consider what the prophet Isaiah wrote about his receiving God-breathed words.

> "Who has believed our message (*report*)? And to whom has the arm of the Lord been revealed" (Isa. 53:1).

In this scripture, Isaiah knew that he was a prophet called by God. He delivered that message to anyone who would listen to it. Since I had an out of body experience that revealed directly to me, "***You are one of the prophets of the book of Revelation!***" I can relate to what Isaiah wrote.

God calls people in the category of prophets to speak or to write of their message. My message is direct. Also, my report confirms that the book of Revelation is spot-on in the twenty-first century. This book carries a profound message for people that will pass into "*the last days*," as explained in the Bible. The book of Revelation is highly credible, whether you accept or reject it[3]!

Now consider the phrase: "... ***one of the prophets of the book of Revelation.***"

Since the book of Revelation is in the sentence, I find this awe-inspiring for two reasons. The author of the book of Revelation was in the inner circle of Jesus' closest followers. John and Jesus were tight. Second, what it means for the planet nowadays.

We can change the phrase "book of Revelation" to "book of the Apocalypse." In general, people view this book as foretelling about a time of great trouble. However, the word "Apocalypse" means to unveil or to uncover. Let us check out a biblical sign foretold in the gospel of Luke related to the Second Coming of Christ.

> "*... and great signs from heaven*" (Luke 21:11b).

Since I had an out of body "spiritual" experience with the sentence. "***You are one of the prophets of the book of Revelation***!" Is that a great sign from heaven that meets the meaning of the recorded text in the gospel of Luke?

In my way of thinking, a great sign from heaven is one that comes from God alone. It is fresh in the end times. My out of body experience about the book of Revelation agrees with that viewpoint. Because it is new and unexpected. In fact, I expect many will think I am crazy!

Another sign found in the gospel of Luke proclaims.

"*... there will be signs in sun and moon and stars*" (Luke 21:25).

To be sure you understand, the hidden period of 14,000 days is time-dependent. We must know the positions of the sun, the moon, and the stars. The burning down of Herod's Temple on the 14,000th day suggests that God stage-managed that event. Also, it reveals that humans are at times under God-like control so that a precise event occurs on an exact day. It seems that is what happened on the day that Herod's Temple burned to the ground.

Now combine these two phrases from the gospel of Luke. We have "*great signs from heaven,*" and "*there will be signs in sun and moon and stars.*"

It appears that the gift of spiritual knowledge may be required. In chapter nine, I will write about the "*Anointed Brain*" experience. You ask, "*What does that mean?*"

Let me give you a hint.

The "*anointed brain*" experience reveals the importance of using logic, historical, and biblical evidence to uncover the repeating 14,000-day cycles. It is truly why I was set apart as a prophet. Even before I was born. I hope you take this statement to mean that this is not about me. I perceive myself only as a vessel filled with the Holy Spirit. God ordained this before my parents conceived me. In truth, God has blessed me.

When you read about this spiritual anointing of my brain, consider my request of you. I strongly suggest that you take the time to read or listen to the next book. Eternal Life - *Why you should expect to live forever*! Because that book holds the substance of my research. This book reveals that we are living in a time that results in a spiritual awakening.

What this shows is that the biblical texts hold information that link together the twenty-first century to the time of Israel in captivity in ancient Egypt. Those are the historical times in which the 14,000-day signs begin. And we are living in a time where those days will end. At the highest importance is that you come to realize that you can expect to live forever. I write that with confidence!

Conclusions

I HAVE PLACED THE "CONCLUSIONS" RELATED TO THIS OUT OF BODY experience about the book of Revelation in chapter seven. That is because chapters six and seven are about the book of Revelation and the out of body experience.

Please read chapter seven, then you will come to the conclusions section.

7

WILL THE APOCALYPSE AFFECT YOU?

THE HOLY SPIRIT BREATHED THIS MESSAGE TO ME IN 1990. "*YOU ARE one of the prophets of the book of Revelation*!"

How many years have passed since the author wrote the book of Revelation? On that day, he wrote about hearing a divine voice.

> "[10]*...I heard behind me a loud voice like the sound of a trumpet,* [11]*saying, "Write in a book what you see..."* (Rev. 1:10-11).

Did the author of the book of Revelation have an out of body experience?

> "[1]*...I looked, and behold, a door standing open in heaven, and the first voice which I had heard, like the sound of a trumpet speaking with me, said, "Come up here, and I will show you what must take place after these things,"* [2]*Immediately I was in the spirit; and*

> *behold, a throne was standing in heaven, and One sitting on the throne...*" (Rev. 4:1-2).

Perhaps the author did have an out of body experience. Because he was in the spirit and appeared before the heavenly throne. At least, it is implied.

Think about the following ideas. Most biblical scholars agree the author wrote the book of Revelation about AD 90 – 96. They typically cite its author as the Apostle John. So, John, who knew Jesus in person, wrote the book of Revelation near the end of the first century.

Then an unknown person living in the late twentieth century receives a message about the book of Revelation! The modern era message comes with two unexpected factors. First, that someone is one of the prophets of the book of Revelation. Second, there is an out of body experience to confirm the spiritual realm to that person. These unforeseen factors are meant to shock you. Are you ready to awaken?

What does this reveal about the book of Revelation, which most people view as cryptic?

To begin, the revelation between two ordinary people separated by about nineteen-centuries means this book is Canonical. We can conclude that time does not control the Source of the book of Revelation, nor does space. The Source of the book of Revelation transcends both time and space. The Source is supernatural.

It is an important book that every person on Earth needs to read and to study.

People on Earth will be in two of the categories below. Think about your groups in the bulleted list below. "What do the book of Revelation and the out of body experience message in the twentieth century mean for each unique audience cited below?"

- Christians that read this book!

- The nations or the entire world!
- Non-Christians, people who do not believe in Jesus, the Messiah as recorded in the gospels!

What Does This Mean for Christian Readers?

Since the book of Revelation is a time of weighty problems, are you ready for the return of Christ?

In my opinion, we probably do not have a lot of time. Yet, it is not up to me to reveal anything about the time of Christ's return. This is because no person knows the day or the hour. However, the signs of Christ's return are all around us. The fact of my out of body experience about the book of Revelation strongly suggests you need to get serious.

Now consider the phrase, "... *one of the prophets*," which suggests that there are other prophets of the book of Revelation. Maybe there are other prophets of the book of Revelation living right now on planet Earth?

Maybe they even know that they are one of the prophets of the book of Revelation?

What is necessary for a prophet of the book of Revelation? The answer is straightforward. If you are a Christ-follower, ponder the text:

"... *the testimony of Jesus is the spirit of prophecy*" (Rev. 19:10b).

If you are a Christ-follower, have you been telling others your personal testimony? Are you telling others about the influence of Jesus Christ and the Holy Spirit in your life?

Be sure, the verse cited above refers to every Christ-follower. To me, this verse expresses the idea that Christians need to share their faith in Christ. That means telling others how Christ has played his part in your personal life. In my view, Christ-followers that share their story are also "***one of the prophets of the book of Revelation.***"

Some of you will say to yourself, "*I do not know or have the strength to do this.*"

You may consider getting involved with this ministry in different ways. You can read the blogs and keep informed to help others gain insight as well. Use your social media account to bring other people to the website or contact information in Appendix B. Also, pray for your friends. Consider the importance of it as I share this personal revelation received by an out of body experience that I am "***one of the prophets of the book of Revelation.***" Would this affect the lives of others?

If a person likes to think or use logic, maybe they would prefer to read book two in the Expect to Live Forever book series. Will the words in Eternal Life - *Why you should expect to live forever*, better impact them?

If you share your testimony with others, then you are telling others that they can have eternal life. All Christians will be victorious over death and hell. In contrast, if you choose to reject Jesus, then......

What if you do not share your faith with others? Who is responsible?

To begin, you must prepare yourself for eternal life. The fact that you are reading this book suggests that God is leading you. Will you become a person who watches from the sidelines?

Or will you get involved as a disciple of Jesus?

Will you help others find eternal life?

Do you think about your friends?

How about your friend's friends?

Consider your enemies who live under spiritual deception. Also the fact that they do not even know they are deceived?

If you take steps to get involved, rest assured that your decision to do so will have a massive impact on the world. Every person on the planet is PRICELESS. It would do well to remember that our enemy is spiritual, not human, in form.

What Does This Mean for the Nations?

On the day that Jesus died on the cross, he stood before Pontius Pilate, the governor of Judea. Pilate questioned Jesus,

---

> 33"*You are the King of the Jews*?"
>
> 34Jesus answered, "*Are you saying this on your own initiative, or did others tell you about Me*?"
>
> 35Pilate answered, "*I am not a Jew, am I? Your own nation and the chief priests delivered You up to me; what have You done*?"
>
> 36Jesus answered, "***My kingdom is not of this world***. *If my kingdom were of this world, then My servants would be fighting, that I might not be delivered up to the Jews; but as it is,* ***My Kingdom is not of this realm***" (John 18:33-36).

At the time of Jesus' death and rising from the dead, Jesus said the kingdom of God was not of this world. In contrast, the texts in the book of Revelation change this idea. A complete overthrow will occur on Earth. Consider the passages at the blowing of the seventh trumpet in the book of Revelation.

> "*The kingdom of the world has become the kingdom of our Lord, and*

of His Christ (Messiah), and He will reign forever and ever" (Rev. 11:15).

If you take the time to read and study the book of Revelation, what will you learn?

You will begin to understand that the final days of humans controlling the political sphere do not end well. In fact, the scriptures reveal that if Jesus does not return, we will destroy the entire planet.

Based on the Apocalypse and my experiences when demons manifested, the goal of the demonic realm is to destroy the Earth. In the satanic world, lying and cheating rule.

Consider the coming destruction based on Jesus' words about his second return to the Earth.

"21 *... there will be a great tribulation, such as has not occurred since the beginning of the world until now, nor ever shall.* 22 *And* ***unless those days had been cut short, no life would have been saved;*** *but for the elect those days shall be cut short*" (Matt. 24:21-22).

These words, as recorded, suggest that humans will destroy the Earth if Jesus does not return. There is also a deeper meaning behind the human struggles. I present a different worldview of the demonic realm in the Psychic Bird Story. (Read: 'Secrets - *never heard until now* - of the Spiritual Realm,' chapter one). In the end, the influence of fallen spiritual beings will bring the world to self-destruction. In contrast, according to Jesus' words in Matthew, He will step in to stop the massive bloodshed.

Perhaps you think this cannot happen.

Then I need to remind you. Humans created the first weapons of mass destruction at the time of World War II. Combine this knowledge with prophetic texts recorded in the book of Revelation and you perceive a bleak future. Please realize that humans already have the potential to destroy the Earth. In fact, humans could make the Earth uninhabitable.

Consider a prophecy about the city of Damascus found in Isaiah 17. The texts foretell of its complete ruin. That it will happen in a single night. These words seem to predict a nuclear weapon that will destroy the city of Damascus. Consider the first and final verses in the foretold ruin of Damascus, recorded about twenty-seven centuries ago. Long before humanity created weapons of mass destruction.

"1... *See, Damascus will no longer be a city but will become a heap of ruins.*

14*In the evening, sudden terror! Before the morning, they are gone*" (Isa. 17:1, 14, NIV).

How would a human being foretell of a nuclear level event more than twenty-seven centuries ago? That Damascus will cease to be a city in a single night! That it will become a heap of ruins, all in a single night! Think about that! It sure is a lot of destruction in the span of a single night based on knowledge from millennia ago.

Also, consider the timing for the obliteration of Damascus by a nuclear weapon. This has only become possible in recent decades. Remember Jesus foretold of the Jewish people returning to their homeland. They regained control of Jerusalem with a 14,000 days alignment to the Feast of Trumpets. This also brings the prophecy about Damascus to the forefront. I must add the sign of the out of body experience about the book of Revelation. When that occurred in 1990, many nations stockpiled nuclear weapons.

What Does This Mean for Non-Christians?

Across history, religions rise and fall. Why?

Because evidence exposes false beliefs. Every flawed belief system will collapse! Is Atheism flawed?

Known science supports that Atheism will also end[1]. If you read the Expect to Live Forever book series, you will learn that known science confronts what Atheism teaches. The problem is so pervasive that Atheist people choose to ignore it, or they want you to believe in their make-believe world.

There are many examples of false belief systems that have risen, then went extinct. Consider the examples of the religions of Egypt at the time of Moses; the pagan religions of Rome, Greece, and Persia at the time of Jesus, and many others.

Do you want to know why I expect other beliefs to collapse?

In the late twentieth century, the Holy Spirit revealed. "***You are one of the prophets of the book of Revelation!***" Then the Holy Spirit validated the spiritual realm with an out of body experience. Also, the repeating 14,000 days sequences that foretell future events on an exact date reveal that God transcends time. God exists in the future and tells us what will happen. The three items I just wrote about support, even validate the Christian faith. Let me bullet these for clarity as follows.

- Prophet of the book of Revelation
- Out of body experience verifies the spiritual realm
- Repeating 14,000 days sequences to future events reveals God exists in the future and tells us.

These bullets summarize that God wants a personal relationship with each of us. That God transcends time. That only the Bible with these prophetic texts has words from beyond space-time. This evidence reveals the Christian faith as having a scientific foundation. Also that its theology is accurate and correct.

In contrast, what does this mean for non-Christian beliefs on planet Earth?

Perhaps those beliefs come from spiritual deceivers! Or those beliefs come from human reason, which includes human error. Or that non-Christian views are based on some variation of the first two sentences in this paragraph.

Flawed Beliefs Summary

We can classify religions into two major categories. First, those that teach that God transcends both time and space. Or those that lecture God is the universe. This approach categorizes beliefs based on the universe alone. It is a very clean separation of major religious views.

The word, Pantheism, is the division of religions that teach God is the universe. In this view, God does not transcend time or space. Based on the third bullet, Pantheism will end in the future. Because those gods do not transcend time. They cannot foretell the future centuries ahead of time. For example, a specific event will happen on an exact date. Also that the final incident reveals spiritual truth. The book of Revelation predicts the end of Pantheistic beliefs.

Then we have the opposing view of God. This view reveals that God transcends time and foretells the future. These are transcendent beliefs. The list of transcendent beliefs is small. You likely know of the word "monotheism." The word "mono" means the word "one." Monotheism teaches there is only one God. These faiths include Judaism, Christianity, Islam, and offshoots such as Bahaism.

But there are vast differences in these beliefs, especially between Christianity and Islam.

The research shows that biblical texts foretell the future accurately. The Bible has Source texts from beyond space and time. However, the

Qur'an has no recorded prophetic words to confirm its source texts came from a Being that transcends time.

Daniel's "time-oriented" prophecy has words spoken from the mouth of the Angel Gabriel. Also, Mohammad, the prophet of Islam, claimed to receive the texts from the angel Gabriel. If that is true, then the Angel Gabriel's words must agree when we compare words in the Bible with passages in the Qur'an.

Daniel's time-based texts foretell of precise events that occur on an exact day in the distant future. Since those texts are from Gabriel, the Qur'an should at least be equal. In contrast, the Qur'an is void of true prophecy.

Even worse, the Qur'anic texts disagree with the Bible. How can we know this to be true?

To uncover this issue, we must compare the writings of these two sourcebooks. Each book must discuss the same subject. This evidence shows the two books contradict one another. Because Jesus is the common subject that the Qur'an and the Bible do not agree with one another.

The Qur'an claims that Jesus did not die on the cross. In fact, the Qur'an claims that Jesus never got on the cross.

In contrast, the central theme of the Bible is that Jesus died on the cross, then rose from the dead.

Which viewpoint is correct?

We know that Daniel's time-based prophecy foretells precise events that occur on an exact day, centuries into the future. Also, that these texts are stage center for Jesus alone. Roman soldiers burned down Herod's Temple on the 14,000th day. We have also learned that this event could only occur from God-like influence on human activities. Because temple destruction events have spiritual meaning. This evidence verifies that Jesus did, in fact, die on the cross and rise from the dead. Because Jesus' death is the final sacrifice. Jesus replaced the temple. The biblical texts call attention to this idea as foreign soldiers burned down both temples on the 14,000th day! This wonder sticks out like a sore thumb.

In contrast, the Qur'an has no texts that foretell of a specific future event that will occur on an exact date. We must stress "centuries into the future!" Also, the final day will have high-level spiritual meaning!

This evidence reveals that the Qur'an is not from a transcendent Being. The research in the Expect to Live Forever book series shows that the Qur'an does not measure up. Because the Qur'an has zero texts that foretell of a precise event that will occur on an exact date, centuries into the future[2].

Other things recorded in the Qur'an and taught by Muslims show that the Qur'an holds flawed passages from a deceiver. If you want to learn about these issues, I record them in the Expect to Live Forever book series.

I have already mentioned the Psychic Bird Story. Everyone in the room watched this demon influenced event. That true story, like a freaky story from the Twilight Zone, teaches that demonic spirits influence people.

If you are not a Christian, I challenge you to open your heart and mind to what I am writing. Do some research by reading, Eternal Life *- Why you should expect to live forever*! This will help you to grasp why this is important. You can also write to us on the contact website listed in Appendix B.

Conclusions

WHAT CAN WE CONCLUDE FROM THIS OUT OF BODY EXPERIENCE ABOUT the book of Revelation?

For sure, the out of body experience about the book of Revelation validates the spiritual realm. We have souls as I felt my spiritual arms and hands the same as my physical parts. Finally, this verifies the book of Revelation is Canonical. This means that only the Bible is the Word of God.

If you were present in the room, you would only have noticed that I was wide awake and how I was holding my physical arms and hands.

It is doubtful you would know about the out of body experience that lasted only three-seconds. You would not have heard the sentence. "***You are one of the prophets of the book of Revelation!***" You would not have listened to the voice of praise either.

However, what you can learn and know is valuable. You can study the repeating 14,000 days sequences. You will discover that the evidence from history supports that the time-based texts did, in fact, foretell the future hundreds of years before. My report of the out of body experience links to this factual evidence.

8

THE VISION OF UNTOLD BILLIONS

In 1990, I worked as a Technical Writer for an aerospace company. I did not have lots of money as I was paying to self-publish my first book entitled, '*Threshold of Eternity*.' Timmy, the guitar player in the Bible study, asked me if we could share the cost of an apartment's rent to save money. I agreed. On the day that I moved to the new apartment, another unexpected event occurred.

Perhaps you recall my writing about the healing experience. I began to feel the presence of God every day since then. When you find yourself in such a deep relationship with God, things can happen that you did not expect. In fact, it could be as simple as lifting your right foot, and something unexpected will begin. Are you ready for another unexpected event?

On Saturday, May 26, 1990, I was moving things from my apartment to a much better place. We finished moving by noon. To wrap it up, we ate at a pizza place. After eating pizza, I drove from that restaurant to my old apartment. During that drive, I listened to Christian radio as they played a song by Christian Musician Dallas Holm. He sang a song called '*Jesus Saves*.' I sang along with Dallas Holm as I drove.

The words to the song are about the '*Untold Millions*' who never get to hear about the good news of Jesus. However I was singing *"Untold BILLIONS"* instead of using the word 'Millions.' I thought Dallas Holm used the word, 'Billions.' I parked in front of the old apartment and finished singing the song on the radio.

I pulled the key from the car and walked into the apartment. I was unprepared for what would happen. Can you imagine seeing billions of people while you are inside an apartment?

After entering the apartment, I closed the front door and walked all the way across the front room into the back bedroom. As I walked back into the living room, I suddenly went into a trance-like state. I entered a vision that overwhelms your physical senses. The scene I came into had billions of people present. I was inside of a prophetic vision that included billions of people.

You may wonder, "*What is it like to be in a vision from God*?"

I found myself looking at the face of a man from Africa. He was only fifteen feet in front of me. He had one knee on the ground and a huge smile. He was delighted. Also, he had ears that pointed straight out, like he could hear my words from miles away.

Next to him was another person. Then I noticed the front row. I was on a stage looking down at the people in the front line. The platform I stood upon was only slightly elevated, with a few steps at the front of the stage. Behind me was a throne. I felt like the presence of God was on the throne behind me.

My eyes started to look at the people behind the front lines. Then I noticed that there were a lot of people in this crowd. As my eyes focused on the people out about a hundred feet, I could see more people going further back.

I looked at the people over a mile away. I noticed there were more people behind them.

I looked to the horizon. It was at least ten miles away. There were people out there too. They were so far away; I could not make out their faces. I just knew that people were that far away in this gathering of people.

The horizon above the people was golden colored with white. It was pleasing and delightful to be there.

At that time in the vision, I could not remember that I had just been singing about untold billions. I could not think. It seems I was meant to see that scene and to report it. I was so immersed that I was just taking in that scene with billions of people that went to the horizon.

As my eyes pulled back toward the front of the people gathered there, I noticed two huge angels. The two angels stood about 150 feet from the front of the stage. They may have been archangels because the people standing around them only reached to their waste. Each archangel was wearing a long, flowing white robe.

Both angels were looking at me. They fastened their eyes on my eyes and held steady. I looked into the eyes of the angel on my right. I noticed that he turned to look at the angel on my left.

Then I focused on the angel on my left. His eyes looking intently at me. For sure, his gaze into my eyes was deep and intense. His face was almost rectangular in shape, with sharp features and golden hair. I watched as that angel picked his left arm up with the white robe hanging below his extended arm. The angel moved the palm of the hand to point up. As the left arm lifted, his eyes remained affixed to mine. As this angel pointed toward the people at the front, it was as if he wanted to show me the people.

I watched the angel begin to turn toward the left. That enormous left hand and arm, sweeping to the left and pointing to the billions in the multitude. After moving his arm to the left about twenty degrees, the archangel began to turn his face to the left to look at the great multitude that stretched to the distant horizon. For sure, billions of people stood before me in that place.

Suddenly, the vision ended!

I found myself walking in circles in the living room of that apartment. As I walked, both of my hands began to flash with the fiery presence of the Holy Spirit. I noticed a spirited flash from my wrist down into my palms. Then this flash moved back into my wrist. Each flash caught my attention. After about ten flashes back and forth in about two-seconds, I stopped walking.

I recalled the vision, fresh and active in my mind. I said to myself, "*I just had a vision*!"

I began to recall this vision. I realized that I had just been inside this vision like I was truly present at that place. Also, when I put everything together, I realized I had been singing about 'untold billions' with Dallas Holm and that this was a vision very similar to what is recorded in Revelation, chapter seven.

After thinking about what had just happened, I realized that I had lost control of my body. I was in another world that you could never imagine.

Seeing a spiritual vision from God is a lot like watching a video play out right in front of you. However, instead of just watching the video, I found myself acting a part inside the video as it happened.

Also, I did not realize I was in a trance until the vision had finished. In this deep trance-like state, I stood in front of billions of people in heaven. Everyone there was kneeling or standing before the throne of God. This is the same vision of John in the book of Revelation. The following graphic gives a clue of this vision.

The Vision of Untold Billions Phenomenon

"[9]*After these things I looked, and behold, a great multitude which no one could count, from every nation and all tribes and peoples and tongues, standing before the throne and before the Lamb, clothed in white robes, and palm branches were in their hands;...*

[13]*Then one of the elders answered, saying to me, "These who are clothed in the white robes, who are they, and where have they come from?"* [14]*I said to him, "My lord, you know." And he said to me, "These are the ones who come out of the great tribulation, and they have washed their robes and made them white in the blood of the Lamb.* [15]*For this reason, they are before the throne of God; and they*

> *serve Him day and night in His temple; and He who sits on the throne will spread His tabernacle over them. [16]They will hunger no longer, nor thirst anymore; nor will the sun beat down on them, nor any heat; [17]for the Lamb in the center of the throne will be their shepherd, and will guide them to springs of the water of life; and God will wipe every tear from their eyes*" (Rev. 7:9, 13-17 NIV).

MY VISION WAS DIFFERENT IN THE FOLLOWING WAYS;

According to verse nine, no one could count the great multitude. However, I knew there were billions of people that stretched to the horizon. That distant horizon consisted of breathtaking colors in shades of gold and white. I state there were billions only because I was singing about 'untold billions' with Dallas Holm. How many billions? I do not know! In this vision, I know they numbered in the billions (see verse nine). I would believe that there were many billions in the vision I saw.

Most of the people in the great throng had white robes, as told in the book of Revelation, except for the people near the front. They were wearing colorful clothes, and I could see their facial features. These people were from the twenty-first century.

However, no one had a palm branch, as described in verse nine. The people whom I saw up close in the vision were delighted and smiling. They were staring at me.

For the people in the distance, I could not see their faces. I could only see that the people stretched to the distant horizon.

Why Would I Have Such a Vision?

To begin, I received the personal revelation that, "***You are one of the prophets of the book of Revelation***!" Then the sudden out of body experience that happened in the twentieth century. The vision is very similar to the texts in Revelation, chapter seven. This vision of billions revealed to me that false belief systems on the Earth will collapse in the last days. In fact, the foundational teaching of the book of Revelation is about the return of Jesus to Earth, along with successful evangelism on a global scale.

The keenest part of my work is the 14,000th day that Romans burned down Herod's Temple. Please realize the prophetic texts do not give us numbers to calculate the time from Palm Sunday to the burning down of Herod's Temple. Regardless, we find exactly 14,000 days hidden in Daniel's time-based texts. Not only do the passages foretell the future, but the words also show divine control of human events. Jesus' death and resurrection replaced the temple. Because Jesus is the final and ultimate sacrifice. This evidence confronts false beliefs. Especially the false teaching of the Qur'an that Jesus did not die on the cross.

Trance-like States in Biblical Literature

Some people may be concerned that I wrote about being in a trance-like state when I received the vision of billions. However, biblical texts reveal that trances are a method used by God for revelation.

Peter had a vision, as recorded in Acts 10:9-10 and references in Acts 11:5:

> "9... *Peter went up on the housetop about the sixth hour to pray.* 10*But he became hungry and was desiring to eat; but while they were making preparations, he fell into a trance*" (Acts 10:9-10).

Paul was also in a trance-like state. According to the text, Paul spoke to Jesus during the trance.

"*...when I returned to Jerusalem and was praying in the temple, that I fell into a trance*" (Acts 22:17).

Conclusions

Can we draw any conclusions from this vision with me in a trance-like state?

I wrote about this event since it links to the book of Revelation. The vision and my calling go together.

If you were present in the room, you would only have noticed me walking in circles. Also, you would have seen that I would not respond to you during the trance-like state. I was in another world.

The only objective evidence that you can learn relates to the fact that Romans burned down Herod's Temple on the 14,000th day. In fact, both Solomon's Temple and Herod's Temple burned down on the 14,000th day relative to the original sacrificial system.

You can study the repeating 14,000 days sequences. You will discover that the evidence from history supports that the time-based texts did, in fact, foretell the future hundreds of years before. My report of this vision related to the book of Revelation links to this factual evidence.

9

ANOINTED BRAIN

SCIENTISTS STATE THAT WE CANNOT FEEL OUR BRAINS. THERE ARE NO pain receptors in our brains. In contrast, when your physical brain gets anointed, you will feel your brain! Of course, this is another Twilight Zone story from the spiritual realm.

When I was writing my first book, God anointed my brain. For sure, this relates directly to the research on the repeating 14,000 days. This incident was way off the beaten path, even for me. I would also say, this incident seems to verify that the research I have done is not from me. To be exact, it appears that the repeating 14,000 days sequences were literally, revealed.

Think of it this way. The 14,000 days sequences were always in the biblical texts. However, we only learn of them at this late date in history. Therefore, logic would suggest that God hid these periods from everyone for thousands of years. If that is true, uncovering these 14,000 days sequences would be due to an act of God.

I learned of the 14,000 days about 1986. Josephus watched the Romans burn down Herod's Temple. Josephus was the only source I knew of in 1986. Yet I was able to find the hidden 14,000 days. The precise date for Romans burning down Herod's Temple is Sunday, August 5, AD 70. Also, you may recall that my father's birth date occurs on August 5.

It seems there is much more going on than we can see with our eyes. The depth of these indicators requires a lot of digging. However, this evidence depends on using logic, historical evidence, and the Bible.

The Event

At the time of this event, I worked as a Technical Writer for an aerospace company. This happened sometime during August 1990. I was preparing to go to work. In my typical day, I took a shower. Then I went into my bedroom and put on my jeans. Then I walked into the kitchen to get some breakfast. As I walked into the kitchen, I was thinking about my research that supports human affairs are under divine guidance at times of biblical revelation. A thought went through my head. I should emphasize that this thought was a revelation spoken to me. Or better said, I doubt that I am the source of this statement. Based on what happened, I view this incident on the same level as the out of body experience.

I thought to myself. "*This is a logical, historical, and biblical thing*!"

Suddenly a spear made of fiery spiritual energy came down into my head. The length of this spear was about three feet long, and about two inches thick. This flaming and spiritual type spear narrowed to a point on its end. As the spear entered my head, it filled my entire brain. Just so you understand, the spear did not pass in and out of my head. That spear entered into the brain and stayed in my brain. It was like the target was my brain. My brain filled with spiritual energy.

I felt every cell in my brain churning with spiritual energy.

The first graphic below shows the spear of spiritual energy about to enter my brain. A unique aspect of this event is that I saw this spear coming toward me before it entered my head. It seems that the earlier anointings made me aware of what would happen. I knew that the spear would enter my brain. If you were in that room, you would not have seen the spear of spiritual energy coming.

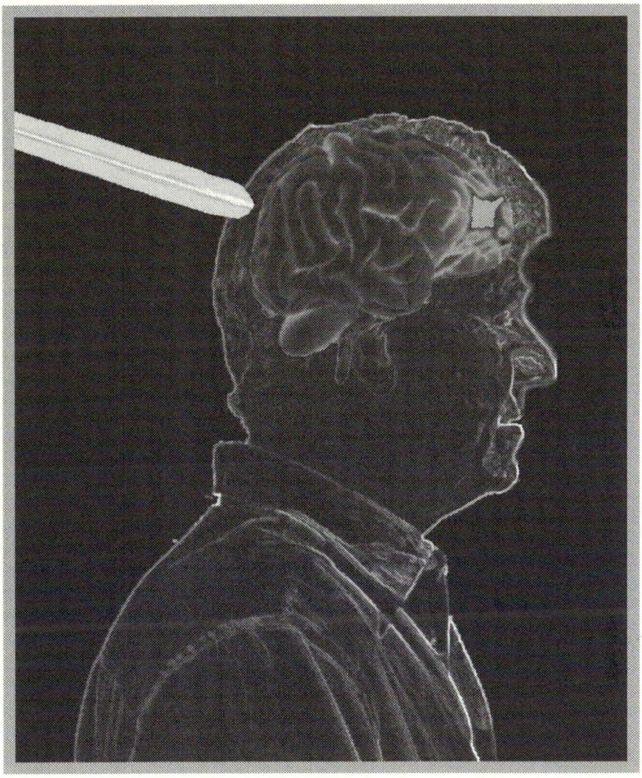

Spiritual Anointing of My Brain Phenomenon

As you are reading about what happened to me, you may think of it as horrifying. However, I was laughing.

The second graphic below is an attempt to reveal just how dynamic my brain felt during this event.

My brain tingled so hard that I held onto the countertop in the kitchen to keep from falling over. As this was going on for about ten-seconds, quite a long time, I said to the Lord out loud. "*If you don't stop this Lord, I am going to fall over!*"

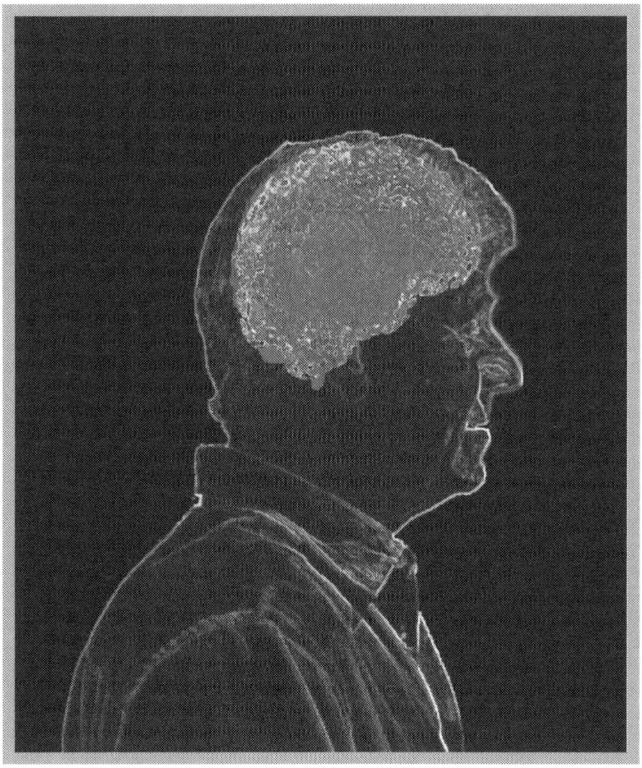

Spiritual Energy Like Fine Granules Churning Phenomenon

I was laughing during the time that I spoke to the Holy Spirit. When I said those words, my brain stopped churning and tingling. Everything was back to normal, instantaneously.

Why did this happen?

What this meant is that the research is not from me. To be clear, God-like help showed me the periods of 14,000 days. At the same time, the study comes from using logic, historical evidence, and the Bible.

Maybe I received the gift of knowledge when my brain got anointed by God! However, you may perceive that I received the gift of knowledge when the ball of power descended on my head in 1973. That anointing is described in chapter three.

No matter how you think of this research, the conclusion is clear. I will bullet the outcome in sequence. I think this is extremely important. I have done this so that logic stands out. You can think about this if you need to.

1. Biblical texts foretell of specific future events.
2. The events will occur on an exact day.
3. A prophet recorded the words on paper centuries before.
4. Destruction of Herod's Temple on the 14,000th day requires God-like control of human events. Also, foreign soldiers burned down Solomon's Temple on the 14,000th day, unveiling repeatability with God-like influence.
5. Burning down both Solomon's Temple and Herod's Temple unveils them as a high-level spiritual incidents. What can we learn and know from this evidence?
6. Jesus' death and supernatural resurrection stand out as replacing the earthly temples. Because Jesus the Messiah is the ultimate sacrifice. The earthly temples were replaced by Jesus, who alone gives us eternal life.

This information becomes relevant in the modern era. People have become distrustful of the Bible. However, the evidence reveals that God transcends time. People who read and study the Bible are on the right path.

It appears that I was born to do this. That I was called to be a prophet of the book of Revelation before I was born. This has little to do with me. In fact, more than forty-six years have passed, and it seems like I have done so little with the materials. However, there may be a divine purpose for the release of this research in 2019 by writing this memoir.

I remember telling the "anointed brain" experience to a friend at the church I attended back in 1990. She did not keep it to herself. Very soon, I heard another single lady say, "*You know, the guy with the anointed brain*!"

Other people were laughing with me!

This anointing of my brain occurred more than six months after the out of body experience recorded in chapter six. Think about a person with an "out of body experience" about the book of Revelation and the anointed brain experience. What does that mean to you?

These events strongly suggest that I received God-like help. Based on this divine anointing of my brain, I would recommend that you read and study the second book in the Expect to Live Forever book series. I recommend that book more than anything else that I write. Please read Eternal Life - *Why you should expect to live forever*! I hope you understand this book holds a wealth of information that will solidify your faith in Christ.

Conclusions

Can we draw any conclusions from this anointing of my brain based on the words? "*This is a logical, historical, and biblical thing!*"

If you were in the room, what would you have witnessed?

To begin, since this was a subjective thought, you would not have heard me say that sentence out loud.

However, you would have seen me grab the kitchen countertop to steady myself. You might have asked me if I was ill. You would have noted that I was holding onto the kitchen countertop for ten-seconds. Also, I was laughing.

Then you would have heard me say aloud during the time I was laughing, "*If you don't stop this, Lord, I am going to fall over*!"

I can only imagine the conversation we would have had after that!

The 14,000 days hidden in the prophetic texts are based on logically discerning the exact dates of those ancient events. Other people besides me uncovered the same precise dates. The fact that other people discovered the same dates reveals that they also used logic and evidence.

The only difference between the others and me is that the Holy Spirit anointed me to find the 14,000 days hidden deep in the biblical texts. Please prepare for yourself, and help others as well.

10

LISTENING TO THE HOLY SPIRIT

I BEGAN STUDYING THE BIBLE WHEN I WAS ONLY A CHILD. IN SECOND grade, I read a ten-volume set of Bible stories. I spent many hours reading and thinking about those remarkable stories.

When I was twenty years old, the Holy Spirit anointed me before preaching a sermon on the Second Coming of Christ. That really got my attention and caused me to think in-depth about my viewpoints of the Bible. That phase ended when the Holy Spirit let me know that I needed to write about my research. As you may recall, that happened in 1987.

After that anointing, I continued with the research and began to write. I did not understand everything that I wrote about. It has taken me decades of thought and improving the study to the point that I can defend it against critics.

In recent years, I have found materials that confirm the 14,000 days. The sequences begin with Moses, then end in the twenty-first century. You may say I have come to the end of this research. But who knows what else we will find as we dig in-depth into the texts of the Bible?

It appears that God speaks and moves the authors of the Bible. My research verifies that the time-based passages foretold the future. That God transcends time. This evidence validates that only the Bible has God-breathed texts.

On August 23, 2019, I had completed a chart on the Elephantine papyri with their precise dates for an Appendix in the book, Beyond - *The Coming Prince by Sir Robert Anderson*. I noticed a problem with one of the science-based dates. As I researched that single problem, I discovered that I had simply typed in the wrong year on the precise time. I corrected and reviewed all the data once again to ensure there were no mistakes.

As I finished that chart and sat back in my chair, the Holy Spirit anointed me again. Both of my ears burned with a fiery presence. After feeling the temperature in both my ears increase at a consistent rate to a soothing and relaxing warmth in each ear, I felt my chest begin to tingle with this fiery presence. After about fifteen-seconds, the burning subsided. This was a pleasing experience.

What is the meaning of the anointing that came upon my ears and chest?

Unless I get specific words, I must interpret what each experience means to me. By particular words, I am referring to the out of body experience with the sentence, "*You are one of the prophets of the book of Revelation*!" Also, I have heard actual words in thought patterns and feel the moving of the Holy Spirit with those words. However, that is very rare.

For me, the burning ears mean I have been listening and following the leading of the Holy Spirit. Also, that God is pleased with this work based on the fiery tingling presence in my chest. I take it as a sign of intimacy and approval. However, the use of logic and evidence is essential.

How would logic apply based on what I worked on when this happened?

Elephantine is a place found in the Nile River in Egypt. It was a military outpost with Jewish soldiers. They served the Persian Empire about 2,500 years ago. The site was unique in that it had a temple, like a temple in Jerusalem.

At this site, a library kept legal records from 2,500 years ago. If you got married, sold land, or such, you hired a scribe to write a legal document. That scribe would write the event with the date on which it occurred. The important thing is that this lets us find the exact day a person did something 2,500 years ago. Especially when the scribe wrote the Egyptian calendar date beside either a Persian or Hebrew calendar date. The author also wrote down the name of the Persian king.

I use this information with science to date events in that era. This leads to the precise day for the beginning of the time-based prophecy. Because the decree to begin counting time happened when King Artaxerxes ruled the Persian Empire. A good number of those legal documents refer to King Artaxerxes. This is the king that Nehemiah refers to in the Bible.

Those documents are vital to this research. Getting the exact dates is essential too. The fact that the Holy Spirit is pleased has to do with finding the 14,000 days sequences. The Elephantine documents let us use logic to find the 14,000 days.

So those ancient records help show the time-based texts foretell the future. The time-based studies about the 14,000 days found in biblical books do predict the future.

You can expect to live forever!

You must understand eternal life is a fact!

Since most of my research has taken decades and there is a lot of depth to this information, I strongly suggest that you read Eternal Life - *Why you should expect to live forever*! That book has the substance of my life research Or, you can listen to the audio format. I believe your life will improve as you gain insight into the idea that ultimate reality is spiritual. You can expect to live forever!

Atheism teaches that you will NOT live forever. Would you like to know why Atheism is wrong?

I document the evidence in Eternal Life - *Why you should expect to live forever*! Today, atheist educators want to change our society. When you read these materials, you will discover the logic, data, and scientific principles that reveal Atheism stands upon flawed assumptions. Science does not confirm Atheism, as many people believe and teach.

Conclusions

CAN WE DRAW ANY CONCLUSIONS FROM THIS ANOINTING OF MY EARS and chest?

The most important thing is that this reveals we can have a personal, intimate, and deep relationship with Jesus. Some people would genuinely like this. Others would abhor the thought of this relationship. That is because they do not perceive or understand the Love of God!

If you were present in the room, you would not have noticed anything that I wrote about my ears and chest. However, you can study what I wrote about the Elephantine papyri. That I use logic and data to scientifically date each document.

A Chronologer is a person that studies when events occurred. I question many things they teach based on data from ancient sources. You will understand only if you read Beyond - *The Coming Prince by Sir Robert Anderson*. I refer to this book as it has the math, charts, graphs, and many appendices to back up what I write.

11

THE PORTAL

WHAT HAPPENS TO CHRISTIANS WHEN THEY DIE?

I told this story to the gathering at my mother's funeral. The story moved people emotionally and truly inspired them.

This is the only story that I am not the original source since it is from my mom. Perchance, this was the only real-life spiritual experience of my mother's entire life. She told this story to several people, and all the accounts are very close to one another. However, I dug the details out of her more than anyone else in the family. She was reluctant to tell people. It is a true story, and in my opinion, the most refreshing story in the entire book. If you fear death, this story will calm your fears and may even cause you to look forward to passing through what I refer to as "the portal."

You may choose to disbelieve what I write in this story. However, I think it is true because I have had thousands of spiritual experiences.

Also, the fact of 14,000 days sequences reveals God is love. Moreover, the 14,000 days sequences point us to the fact that God transcends time. You, too, can have eternal life. Appendix A for people who decide that they want to make that vital decision. I hope you genuinely enjoy this breathtaking story.

This story occurred during a cold December morning in 1982. My grandmother was seventy-nine years old. She died of an aortic aneurism. To help keep track of the characters in this story, we fondly and aptly called my grandmother, Granny.

Around five in the morning, my mother received a phone call from Granny. She said to my mother, "*Please come quick. I don't feel good*!"

My mom put on her nightgown and drove rapidly to the home where Granny lived, which was only two miles away. When my mom walked toward the house, she could see through the window that Granny never hung up the wall phone.

My mother went into the house and found Granny lying on the floor. Mom kneeled and felt for the pulse, which was not there. Then she lifted both hands up to the level of her head and spread them outward from her face and said, "*Oh my God, she is dead*!"

At that instant, my mother's eyes opened to the spiritual realm. She saw her mother's soul float up out of her body. Also, there was a thin film about six feet above the floor. As Granny's soul passed through the film, that part of her soul would disappear from my mother's sight. It was as if the film was the limit of how much my mother could see of the spiritual realm with her eyes.

As Granny's feet passed through the film, that film popped as the visual part of the spiritual realm ended. However, at the point that the feet passed through the film, and it popped, my mother said to me, "*I felt her joy*!"

There was much more to this unveiling than what my mom saw with her eyes. She also felt Granny's joy as she passed through the portal into the heavens with eternal life. Because Jesus has made way for everyone, who will place their trust in him.

Conclusions

CAN WE DRAW ANY CONCLUSIONS FROM THIS STORY OF GRANNY entering heaven?

We have two things that go together. I had an "out of body experience" related to the book of Revelation. Also, my mother saw her mom's soul pass into heaven along with the emotion of joy. These stories reveal a consistent viewpoint associated with the spiritual realm. Are you beginning to conclude that, "*You can expect to live forever*!"

When your physical body comes to its end, you will also find yourself in the spiritual realm, literally.

I have often wondered what Atheist people think when their souls become obvious to them upon the death of their physical bodies. Think about Stephen Hawking. He came to believe he could explain the universe with science. I must add, Hawking had to make assumptions that are certain to include human error. Since I worked in an engineering environment with engineers and scientists, I sometimes saw their weak assumptions.

When Stephen Hawking died as an atheist, I wonder what he thought as he became aware of his soul?

If you were present with my mother in that room where Granny died, you would have noticed my mom feel for Granny's pulse. Then the response of my mom raising her hands and saying, "*My God, she is dead*!"

It is unlikely that you would have seen the vision or felt the joy that my mother experienced. At least that is our assumption.

This is why I use science for the only "time-oriented" prophecy in the Bible. Because it gives us objective evidence that you can use in your personal quest for eternal truth. You can study what I wrote and find evidence for the spiritual realm.

Let us think about the Atheist people that have their false paradigm. I expect them to disagree. Here is why.

Did you know that some Atheists study the Bible. They want to prove it to be wrong. When a book in the Bible has miracles, then they declare the book to be untrustworthy.

Since the book of Daniel has fulfilled prophecy, the Atheist people claim it was composed in 165 BC[1]. Why?

Because we all know that no human can foretell the future! I agree that humans cannot predict the future. Only a Creator that transcends time can foretell the future.

The repeating 14,000 days sequences confront the Atheist outlook. The Atheist worldview is "believed" to be based on science. However, the use of science is what brought these time sequences to light. Science, the Atheists' standard for belief, reveals the reality of a transcendent Being behind the 14,000 days repeating sequences.

APPENDIX A

HOW TO START THE ETERNAL LIFE JOURNEY

Did you decide to follow Jesus?

Read only if you want to receive eternal life! When you die, do you want to feel joy as you go through the portal into the Kingdom of God?

> Jesus said, "*No one can see the kingdom of God unless he is Born Again*" (John 3:3 NIV).

To begin, no person on planet Earth can put into practice a lifestyle that pleases God. We fall short of that high mark of perfection. In contrast, Jesus' death and rising from the dead sets us free from reaching perfection. The passages of the New Testament reveal that Jesus alone gives you this faultlessness. To obtain this pure state of righteousness, you place your trust in Jesus alone. As a result, your sin and blemishes go onto Jesus on the day that he died on the cross.

That means every wrong you have done will get removed by Jesus' death on the cross. You are set free and made perfect by Jesus. As a result, you begin a relationship with the Holy Spirit. You receive this salvation that gives you eternal life. You become a new creation and take your first steps into everlasting life.

Maybe you have ideas that do not agree with what I just wrote. At this point, I want to help you understand what I just wrote. Do you think you could never become good enough to become a Christian.

In truth, that is the way humans think. It is also the evidence related to all human reasoned beliefs, especially churches with legal codes you must abide with.

Consider King David in the Bible. He had an affair with Bathsheba and got her pregnant. Then he made sure that her husband died on the battlefield. Those sins are evil. However, God's grace and forgiveness changed King David. Long before King David did these sinful things, the God that transcends time chose him. This means that God knew of David's future immoral and evil decisions. But God anointed David anyway. I write so you can understand that God loves you, despite your weighty sin.

Before reading more, let me explain the God that transcends the universe. He knows more about you than you know of yourself. Every thought you have, whether good, bad, ugly, or beautiful, is known, across your entire life, long before you were conceived. Before the universe began, He knew you. Are you ready to start a relationship with this Being, who loves you more than you could imagine?

In contrast, non-Christian beliefs teach that you attain uprightness by doing good works. In general, non-Christians hope that their good works will be better than their sins. However, Jesus' death on the cross on the Passover as the Lamb of God challenges this view. Because no human can live good enough to ever achieve eternal life. Consider this verse!

"...all have sinned and fall short of the glory of God" (Rom. 3:23 NIV).

This passage reveals that you are not able to obtain eternal life on your own. Take a deep breath and relax. Because what I wrote in the initial three paragraphs is true. The New Testament texts guarantee you are freely given the righteousness of God by placing your trust in Jesus alone. Jesus is the only person capable of securing your eternal salvation, despite just one sin or all sin in your life. For example, regardless of past sins, the sins you will commit today, and all future sins. Some people will not like what I just wrote. Think about King David and the future crimes he committed. Then read this verse.

"...where sin increased, grace abounded all the more" (Rom. 5:20b NIV).

This verse means that despite yourself and all your sins, God favors you! Even if you are not able to accept this grace, God is merciful to you. Why would that be?

"19God was reconciling the world to himself in Christ (Messiah), *not counting men's sins against them. .... 21God made him who had no sin* (Jesus on the cross) *to be sin for us, so that in him we might become the righteousness of God"* (2nd Cor. 5:19, 21 NIV).

So, Jesus has taken all your personal sins upon himself. That is the blame and just punishment for your own sin. If you accept this and place your trust (*faith*) in Jesus for your righteousness, then you instantly become perfect and holy! This happens regardless of yourself and your fallen human nature. The following verse gives a requirement that you believe in Jesus' death on the cross and then rising from the dead.

"9...if you confess with your mouth Jesus as Lord, and believe in your heart that God raised Him from the dead, you will be saved;

> [10]*for with the heart a person believes, resulting in righteousness, and with the mouth he confesses, resulting in salvation*" (Rom. 10:9-10 NIV).

Do you believe what these verses proclaim?

Maybe you will ask, due to your shortcomings, "*Am I guaranteed to be forgiven*?"

> *"If we confess our sins, he is faithful and just and will forgive us our sins and purify us from all unrighteousness"* (1st John 1:9 NIV).

Then say the following prayer with your whole mind, heart, and soul.

> *"Lord Jesus, I need you. I want to be born again and to receive your Holy Spirit into my life. Thank you for dying on the cross for my sins, then rising from the dead. I open the door of my life and receive You as my Savior and Lord. Thank you for forgiving my sins and giving me eternal life. Take control of the throne of my life. Make me the kind of person you want me to be. In the name of Jesus Christ, the Son of God, Amen!"*

Welcome to the family of God and for placing your trust in Jesus, the Messiah, who alone makes you righteous.

To keep informed and up-to-date, please become a member of the website listed in Appendix B.

For sure, you can read the blogs. Also, you can use your influence (*leadership*) on social media or just by telling others to bring them eternal life in Christ alone. Also, find a Bible-believing church and become close friends with your brothers and sisters in Christ. May there be manifold blessings upon your life.

Learn to ask the Lord through prayer for help so you can become Christ-like based on how you talk, how you treat people, and your lifestyle. Read and study Galatians chapter five, Rom. 8:28-29, John

6:37-40, as well as Rev. 22:11-15. Even if you fall short, never give up based on 1st John 1:7-2:2. Just as important, always find time to be with your fellow believers based on Heb. 10:24-25 and First Thess. 5:11. Please study these texts to increase your faith and to grow resilient faith. I strongly suggest that you get involved with an evangelical community of believers. Please sign up for blogs and other changes as this ministry grows. The websites listed in Appendix B will be updated with ongoing changes as we progress.

If you have any questions, please let us know by using the contact page on the website.

APPENDIX B

CONTACT INFORMATION

Contact and/or Get Involved

To keep informed and up-to-date, please become a member of the website.

http://www.harvardhouse.com

http://www.expect-to-live-forever.com

http://www.expecttoliveforever.com

Global-wide Christian Revival

Do you believe there will be a global-wide Christian revival?

Then please go to the Expect to Live Forever website to discover the methods for mass evangelism that will stop Islamic growth, then lead to its implosion. Evidence in support of this viewpoint may be read in chapter twenty, subheading: **How to use this information**, recorded in book two of the Expect to Life Forever book series. Or you will find this in book three, chapter twenty-two subheading in the Expect to Live Forever book series.

ACKNOWLEDGMENT

I am grateful to Sir Robert Anderson (1841 – 1918). He authored The Coming Prince, published in 1894. He laid a solid foundation that is corroborated by ancient astronomy. Then I take his research to new heights.

I am thankful for Pastor Frank, who cheered and guided me as I wrote these materials. He counseled me before I began to write this book.

I am also grateful for my brothers and sisters, their critiques, and influence on the content. I am pleased with Tom, Tocee, Gary, Gilbert, Nida, Sandra, Jeni, Jon, Pastor Charlie, Pastor Eric, Pastor Vincent, Bishop Mohamed, Nancy, and Chad. Each gave me encouragement, as well as support and guidance in writing.

ABOUT THE AUTHOR

John Zachary

Author of Inspirational Christian Mysteries with degrees in Math, History, and Education. His work skills include Technical Writer and aerospace Reliability Engineer. He uses these earthly skills to analyze texts that many believe foretell the distant future. If that is true, we unveil evidence of a transcendent Being. Can we expect, without question, to live forever?

In contrast with his earthly skills, he writes mysteries linked to many thousands of spiritual experiences that began at twenty years of age. The most profound incident is a secret about the book of Revelation that he kept to himself for three decades.

John is married with a son, daughter, and grandson. They love to go places and have fun together.

https://expect-to-live-forever.com

EXPECT TO LIVE FOREVER

THE BOOK SERIES

1. Secrets – *never heard until now* – of the Book of Revelation
2. Eternal Life - *Why you should expect to live forever!*
3. Beyond - *The Coming Prince by Sir Robert Anderson*
4. Secrets – *never heard until now* - of the Spiritual Realm
5. More books with cutting edge material are coming! To keep up to date, consider getting on the email list by going to the website contacts in Appendix B.

I look forward to hearing from you. You can pre-order at Amazon. Or get on email for new releases as well as ministry opportunities.

www.expect-to-live-forever.com

PRE-ORDER — SAMPLE CHAPTER

Book #2: Expect to Live Forever book series

Eternal Life - *Why you should expect to live forever*!

Can We Know That God Exists

We live in a time when all we hear is that religion and science do not get along. At the extreme level, so-called scientific atheists claim that God does not exist!

Consider this fundamental question.

Isn't there any science-based data to prove that God is dead?

The correct answer is no. If that data were available, every scientist would be an atheist!

Perhaps you are thinking, "*Just show me an example of a world-class scientist that is Christian?*"

Consider Dr. John Polkinghorne, who is a theoretical physicist and Anglican priest. Why would a world-class scientist become an Anglican priest?

Before deciding to become an Anglican priest, Dr. Polkinghorne played a role in finding the quark[1]. He has authored five books on physics as well as another twenty-six books on the relationship between science and faith. You may be surprised to know that he has a strong belief in Jesus rising from the dead!

Why would a world-class scientist believe in God and miracles?

Brilliant people will continue to believe in God. In fact, as you read into this book, you will learn that Atheism logically leads to a supernatural realm. Why?

To validate your worldview, you must show how the source of the universe is eternal. However, Atheism only accepts the natural realm. This means that the natural world must be endless, but "known science" reveals that every natural thing will not last forever. Atheism cannot solve this problem. In fact, Atheism relies on unbelievable amounts of blind faith in the impossible!

In contrast, the supernatural realm will never stop working. In this book, we will see how known science and logic unveil the spiritual realm. In simple language, something greater than our universe is required to explain the world and the fact that the cosmos is dying. I will give more details on this in later chapters.

This chapter introduces you to three methods that unveil the spiritual world. Before I write about that, it is crucial to grasp that we must have data of some type. Because data lets us know that the spiritual world is a fact, just like the air we breathe!

Some people may not like using the word "fact" for the spiritual realm. Another way of saying this would be that the data suggests the spiritual realm is factual. We study the data to find out if we can have confidence in our conclusions. If we have confidence in the spiritual realm, then we must also answer the question. "*What is faith?*"

Method One: Texts that Foretell the Future

Do you think it is possible to write words on paper that will predict the future? Not the events that will take place in the next hour or the next minute, but centuries later. For example, "What will happen in three hundred years? Also, what will be the military action on that day?"

Of course, this is not possible for beings such as ourselves. Or aliens in spaceships, if they exist! Also, it is not viable for unseen spiritual beings such as angels or demons! For sure, not even the Archangel Gabriel would know the future as some may think! Because time and space do not let created beings know the future. Creatures, in contrast with the Creator, do not have access to events in the distant future. Time controls everything that created entities do.

Only the Creator can foretell the future. In this book, we find the details that strongly support, even verify, that "time-based" texts predicted specific events that happened on a precise date, centuries into the future.

Moreover, we discover that foretold events have unique ***high-level spiritual meanings***. These events teach us about the spiritual realm and require the event to occur on an exact date, centuries into the future. A summary would mean at least four things about this transcendent Being.

- This Being transcends time!
- This Being exists outside of our universe, where time and space mean nothing.
- This Being only foretells of events with high-level spiritual meaning!

NOTE: In general, all other events do not have God-like control. You have free will to live as you please.

- To cause the specific event foretold to happen on an exact date, centuries into the future, this Being will control humans to bring about that prophesied event on an exact day. I must emphasize that only events with high-level spiritual meaning appear to be controlled at an elevated level! The dates of these foretold events strongly support this viewpoint. This evidence unveils the spiritual realm, like the air we breathe!

As a result, we learn priceless information about ourselves. We can logically conclude, "*We can expect to live forever*!"

What else can we find from words that foretell the distant future?

This method also lets us analyze differing beliefs. Are the differing "beliefs" true or false?

This approach exposes the shortcomings of what so-called scientific atheists believe[2].

Method Two: Observation of the Effects of Spiritual Beings

Most people would wonder if it is possible to view events "caused by" or "influenced by" a spiritual being. The notion, of course, seems hard to believe, but it is possible. How?

Because I was one of five people who saw such an event. I am talking about the *Psychic Bird Story*! This story is recorded in book four in the Expect to Live Forever book series, Secrets - *never heard until now* - of the Spiritual Realm. The five of us viewed the effects of a spiritual being during a Bible study. The demonic spirit made a parakeet in a cage squawk two to three times each second. Also, the bird beat its wings as fast as it could. This spiritually influenced event took about five minutes. The purpose was to distract from my teaching about genuine psychics being possessed by such beings.

Based on that, we can conclude that a spiritual realm exists. The evidence is sound. We can calculate our confidence level for the spiritual realm. There is only one chance out of 10,000 that the bird decided to do this while I taught about demons. This is based on the observation of five people. We all saw it and heard it happen. I personally observed people being distracted by the bird! I suggest you read the *Psychic Bird Story* recorded in book four of the Expect to Live Forever book series.

Consider this question: "*Are there other stories where a group of two or more people observed the effects of a spiritual being*?"

If you read all the books in this Expect to Live Forever book series, you will be able to grasp more about the spiritual realm. This means you need to prepare yourself, so you do not get deceived by religious deceivers or by Atheists, who lack understanding.

I am sure there are other meetings where people have seen the effects of spiritual beings. In fact, I would expect to find more evidence!

Finally, this method is validated by scientific dating specific foretold events to occur on an exact date, centuries into the future. Therefore, two examples are included in the Expect to Live Forever book series.

Method Three: Scientific Principles and Data Reveal Spiritual Truth

Will the universe last forever?

It is a fact that the sun will eventually stop producing light and heat. Because there are limits for the amount of atomic energy in the sun. As a result, the sun, at the end of its life, will become colder than ice. Many stars have already used up their atomic energy. Then they cool off over eons of time. This will happen to every heavenly light in the cosmos.

Scientists[3] came to this conclusion in the 1850s. Since then, science has continued to prove this claim. Every scientist knows that the universe is doomed. It has an expiration date. NASA scientist Michael Pauken[4] writes:

> *"I hope you don't get depressed when I tell you that the universe is going to end one day because of the 'increase in entropy' principle."*

This NASA scientist validates that the universe cannot take care of itself. People agree with Michael Pauken's statement because it is based on known science with vast amounts of data. Therefore, it makes sense to ask:

> "*What does this science-based information reveal about finding spiritual truth*?"

BIBLIOGRAPHY

1.Archer, Gleason L., Encyclopedia of Bible Difficulties, published 1982, is the source of the 530 BC composition date.

2.Jack Finegan, Handbook of Biblical Chronology Table 51, p. 108, (c) 1998.

3.http://numerical.recipes/julian.html

4.A Julian Day and Civil Date Calculator

5.Archer, Gleason L., Encyclopedia of Bible Difficulties, published 1982, is the source of the 530 BC composition date.

6.Archer, Gleason L., Encyclopedia of Bible Difficulties, p283

7.Anderson, Sir Robert, The Coming Prince, 1894

Pauken, Michael, Thermodynamics for Dummies

CAN YOU HELP?

Thank You For Reading Book One in the Expect to Life Forever book series!

I really appreciate all your feedback. I would love to hear what you have to say. Would other people appreciate your insights into this unique and insightful book?

Please leave me an honest review on Amazon, letting me know what you thought of this book. I truly appreciate you taking time to help!

Thanks so much! Many blessings upon you! I pray and hope that you enter the world of miracles and dreams come true, forever!

John Zachary

NOTES

1. The Ultimate Secret

1. This is a statistically significant event, which strongly suggests a factor that influences the outcome. In this case, evidence of transcendent knowledge.
2. A foundational reason to be pro-life. Think twice before voting for politicians that support abortion, or political parties that accept and promote abortion.
3. The Apostle Paul records an out of body experience in 2nd Corinthians 12, through which he received divine revelation. So, it can be a Christian experience that is the will of God.
4. I reveal how this "tingling presence" in my hands became a typical "everyday" occurrence in my life as written in chapter five.
5. What is the difference between an out of body experience (OOBE) versus a near-death experience (NDE)? Cognition is the most significant difference as I was completely awake. I knew the difference between my spirit arms versus my physical arms. Also, my spirit brain did not separate from my physical brain. This OOBE supports that the mind and brain are both spiritual and physical. In contrast, in an NDE, the person is not cognizant, and their reports are from subconscious levels. Therefore, much of what people perceive of NDEs appears to be questionable.

   Moreover, the viewpoints of people who experience NDEs is contradictory as some are Christian, others are from those who believe in reincarnation. If NDEs are correct, then I would expect people with NDEs to only report one of these viewpoints. This is an essential point because the two beliefs are in different categories of religion (transcendent in contrast with pantheism).

   Also, I back up my OOBE in Beyond - *The Coming Prince by Sir Robert Anderson* as it is based on astronomy, ancient texts about the Hebrew/Persian calendars, logic, and scientific data that leads to the premise: "The natural universe cannot exist without a supernatural source." Therefore, we logically expect to see supernatural phenomenon (*miracles*) in this universe. Moreover, scientific dating the only "time-oriented" prophecy with complete accuracy validates that the Source of the texts is a Being, who exists outside the controls of space-time and is not subject to natural laws. Explaining the "time-oriented" prophecy with scientific dating is another dataset that supports supernatural phenomena.
6. The book of Revelation alludes to the Old Testament books without quoting their source texts. Various scholars have detected source texts in the books of Genesis, Leviticus, Psalms, Isaiah, Ezekiel, Zechariah, Deuteronomy, Joshua, Exodus, Daniel, and Hosea.

7. The other two are Peter and James, who was the brother of John, the author of the book of Revelation.
8. In the transfiguration story, Jesus takes the inner circle of three up a mountain where he transfigures in the brilliant light of the spiritual realm. This terrifies the three disciples due to the brightness of seeing the glory of God. You can read this story in Matt. 17:1-8, Mark 9:2-8, Luke 9:28-36, referred to in 2nd Pet. 1:16-18, and likely in John 1:14 "We beheld his glory!"

   Some Atheist scholars teach that the belief in Jesus as the Son of God and second member of the Trinity is the result of Pagans becoming Christians in the final decades of the first century. These scholars typically do not refer to the synoptic gospels but only the gospel of John. This non-biblical viewpoint chooses to ignore the Synoptic gospel stories of the transfiguration found in all the gospels since this story portrays Jesus as being equal to an Eternal Being. Also, they ignore Old Testament texts that support the deity of the Messiah. For examples, refer to Ps. 2:7, Ps. 110:1, Prov. 30:4, Dan. 7:13-14, Mic. 5:2, Isa. 9:6 and Jer. 23:5-6 as biblical texts foretelling of the person of Messiah depicted as a divine human being. Jeremiah's words give us the most intriguing Old Testament texts about the deity of Messiah. This text uses the Tetragrammaton [a four-letter word YHVH or YHWH in Hebrew that Jewish believers will never pronounce because it is the name of G_d and is too holy for humans to pronounce] as the name of the Messiah. Jeremiah's words strongly support the Messiah is a divine character, which is described in New Testament texts.
9. Christians who believe that the book of Revelation was fulfilled during the days of the early day Christians at the time of the Roman Empire destroying the nation of Israel are referred to as Preterists. My out of body experience strongly suggests that the Preterist interpretation is not valid.
10. Jesus revealed to his close followers that the only way to eternal life was through himself as recorded in John 14:6. My out of body experience about the book of Revelation supports Jesus' statement and confronts the teachings of all other belief systems. For a much clearer understanding, please read the evidence presented in Eternal Life - *Why you should expect to live forever*, or listen to the audio version. If you want to get the supporting data to confirm the research, then please read Beyond - *The Coming Prince by Sir Robert Anderson*.
11. I call this concept factual based on the out of body experience about the book of Revelation. Moreover, many of the other experiences recorded in this book support this viewpoint. Some of the experiences are objective as there were other people present as eyewitnesses of a spiritually driven event, especially the events recorded in chapters three, five and nine. I write the most objective experience known as the Psychic Bird story recorded in book four of the Expect to Live Forever book series.

2. Awe-Inspiring and Elegant Words

1. The all-knowing Supreme Being of the Bible knows every event, every thought, and every action before the universe was created. We can deduce from Gen. 1:1 that the Eternal exists in a realm where "time" does not exist, "space" does not exist, and "matter" does not exist. The habitation of the Eternal is entirely spiritual. No one can ever see that universe. We can only know of it by faith. However, my research material support the probability of the spiritual realm approximating 10^(102) at a confidence level of 99.99%. If you don't understand these values, it can easily be stated another way. "*You can bet on it!*"
2. The word "revelation" means to uncover or to unveil divine information of what had been previously and totally hidden or only vaguely known. In the written texts of the Bible, this means that God reveals supernatural information of himself, his purposes and plans, that were guided by the Holy Spirit and delivered through a prophet of God for humanity to consider.
3. The "Eternal Life" book is an easy read without math. If you want to get a book with supporting data to confirm the research, then please read Beyond - *The Coming Prince by Sir Robert Anderson.* Sir Robert Anderson published the Christian classic, The Coming Prince in 1894. My research gives evidence that corroborates Sir Robert Anderson and enhances his work by revealing the 14,000 days sequences that end in the twenty-first century. However there is more to come based on evidence in the Expect to Live Forever book series.
4. Knowledge is about the past. Faith is about the present and the future. I limit my research to historical information using scientific dating, logic, and other methods for events of the past. I do NOT make any future predictions in the Expect to Live Forever book series as that requires faith. My research unveils spiritual knowledge due to vast evidence from history that encroaches on the modern era.
5. The hidden period of 14,000 days in Daniel's seventy-weeks prophecy support that the texts foretell the future. This negates the 165 BC date held by rationalists. Conservative scholar, Gleason L. Archer, Encyclopedia of Bible Difficulties, published 1982, is the source of the 530 BC composition date.
6. Many modern era scholars reject Sir Robert Anderson's research based on astronomical data and their presuppositions. In recent years, I have uncovered evidence that corroborates Sir Robert Anderson's analysis and conclusions. I have placed that information in Beyond - *The Coming Prince by Sir Robert Anderson.*
7. Reference: Handbook of Biblical Chronology by Jack Finegan, Table 51, p. 108, (c) 1998.
8. The website: http://numerical.recipes/julian.html is a Julian Day and Civil Date Calculator, which gives excellent results for ancient events. If you want to verify the dates as well as the information I am writing about, you can use this calculator to do so. You can also hand calculate if you want.
9. The correct "word" for understanding the God of the Bible is "transcendent." This word means that this Being exists outside the natural universe. God tran-

scends time and space. A transcendent Being knows everything and lives in the past, the present, and the future. Foretelling future events to an exact date, centuries ahead of time would be easy. In Contrast, Pantheistic Religions do not have prophecy because their gods are controlled by time and space.

10. This word "transcendent" means that this Being exists outside the natural universe. God transcends time and space. Foretelling future events to an exact date, centuries ahead of time would be easy. In contrast, the Qur'an does not have this required evidence in its texts. This means the source of the writings recorded in the Qur'an exists under the control of space and time. Allah, in the Qur'an, is not transcendent. Muslims live by blind faith with absolutely no evidence that Allah exists. Islamic theology referred to as "Absolute monotheism" supports this last statement. Allah cannot be known or proven to exist. In fact, Islam is a form of agnosticism.
11. The "Eternal Life" book is an easy read without math. If you want to get a book with supporting data to confirm the research, then please read Beyond - *The Coming Prince by Sir Robert Anderson.* Sir Robert Anderson published in 1894, the Christian classic, The Coming Prince. My research gives evidence that corroborates Sir Robert Anderson and enhances his work by revealing the 14,000 days sequences that end in the twenty-first century. But there is more to come based on evidence in the Expect to Live Forever book series.
12. Revelation 15:3 appears to have relevance to the 14,000 days due to the characters of Moses and the Lamb of God (Jesus).

    As you read more of this research, you will learn that Old Testament 14,000 days periods are mystical. However, New Testament periods of 14,000 days are actual days. These 14,000 days sequences unveil a spiritual temple that is not of this world. The book of Revelation gives insight into a supernatural universe wherein its inhabitants will live forever.

    What is written in Revelation 15:3 illustrates the viewpoint recorded in a supernatural universe wherein those inhabitants will live forever.

    Revelation 15:5 refers to the temple in heaven, which the book of Hebrews refers to as a place "not of this world!"

    Revelation 21:22 supports that burning down Herod's Temple on the 14,000th day as an act of God. To gain insight, please read and study Hebrews 9:1 - 10:39.

    Administering the seven bowls of wrath come from the temple in heaven according to Revelation 16.

    I suggest you use a concordance to find the word "temple" in the New Testament and find how it relates to end time events. I ask that you do this knowing the 14,000th day destruction of the earthly temples were acts of God.

3. Anoint!

1. I am sure that many readers will conclude that I was in a Pentecostal or Charismatic Church with speaking in tongues and all types of spiritual phenomenon. However, the church I grew up in and preached in was not of that genre. Regardless of the church setting, I have always been taken by complete surprise time

after time with the leading of the Holy Spirit. I have always been in awe of the love of the Holy Spirit and the breathtaking influence over my life, despite my flawed Christian walk. I have come to believe in an effortless faith. Just place your trust (literally "faith") in Christ Jesus, and everything is going to be all right. I have no confidence in my so-called "good works" for salvation as I can never be good enough. Neither can you be good enough. The good works are just evidence to others that I am indeed a man of faith in the person of Jesus, the Messiah, whose death and resurrection make it possible for each of us to receive eternal life. If you want to join in this life with Christ, follow the instructions in Appendix A to receive eternal life by placing your faith in Jesus, the Christ.

2. No spiritually driven events occurred again in my life until I had completed the research that took me fourteen years. When the next event occurred, I would again be taken by complete surprise.

4. My Call to Write

1. Another reason I have waited to publish this memoir is that scholars reject AD 32 as the year of Jesus' crucifixion based on astronomy. However, since I had the out of body experience on February 1, 1990, I did not give up on finding evidence that would validate AD 32 as the year of the crucifixion. I finally found the evidence written in the Mishnah with supporting data from Babylonian astronomical texts as well as the Elephantine papyri. These ancient sources reveal flaws in the presuppositions held by scholars. The references and calculations are recorded in Beyond - *The Coming Prince by Sir Robert Anderson.*
2. "... the people of the prince who is to come will destroy the city (*Jerusalem*) and the sanctuary (*meaning the temple in Jerusalem*)" (Dan. 9:26b NASB).
3. Wars of the Jews, 6.4.5 https://www.sacred-texts.com/jud/josephus/war-6.htm
4. Reference: Handbook of Biblical Chronology by Jack Finegan, Table 51, p. 108, (c) 1998.
5. Rabbi Akiva, born AD 50, was twenty years old on the day Romans burned down Herod's Temple. Rabbi Yose, a student of Rabbi Akiva, recorded that both Solomon's Temple and Herod's Temple were burned down on a Sunday. Rabbi Yose got this information from Rabbi Akiva.

   Reference: Handbook of Biblical Chronology by Jack Finegan, Table 51, p. 108, (c) 1998.

5. "I Want to See the Power of God"

1. Did this reality catch your attention? If so, I suggest that you study Matthew 13:44. A single verse conveys the value of the Kingdom of Christ!
2. My perception of reality is that the natural universe is entangled with the supernatural. My natural body will die. However, my soul is eternal, and it will never die. My natural body is entangled with my supernatural soul. It is extremely important that you gain insight into the fact that ultimate reality is spiritual. A

deeper understanding of our existence may be gained by reading book three in the Expect to Live Forever book series, *Beyond - The Coming Prince by Sir Robert Anderson*. Or listen to the audio, *Eternal Life - Why you should expect to live forever*!

3. Quantum entanglement tests verify that subatomic particles linked together as a unit can be light years apart, but still be a single unit. When one of the particles is disturbed, the other instantly knows that. The distance between the particles is irrelevant because they are one unit. If you read John 14, you can begin to understand how quantum entanglement helps to explain the Trinity.

6. Prophetic Calling

1. Chapter eleven, "The Portal" documents the story of my mother's eyes opening to the spiritual realm at the death of her mother. It is normal to fear death. In "The Portal" story, you can find comfort as your physical life will someday end. Knowing Christ means there is no need to fear death. I highly recommend that you read chapter eleven.
2. Biblical Reference: Ezek. 33:2-6 has similar ideas. In this reference, Ezekiel uses ideas related to a watchman, who is essentially, a security guard. Think of a prophet as a security guard, sent to warn you of coming tragedy!
3. You should expect that my out of body experience makes my position necessary!

7. Will the Apocalypse Affect You?

1. I present Atheism's flaws in Eternal Life - *Why you should expect to live forever* or Beyond - *The Coming Prince by Sir Robert Anderson.*
2. Muslims will claim there is a successful prophecy in the Qur'an about the Romans defeating the Persians. However, the Qur'anic texts were not recorded on paper until after the events occurred. A true prophecy must be logged on paper long before the event occurs. Logical people conclude the Qur'an has no true prophetic texts equal in the least to the "time-oriented" texts of Daniel.

11. The Portal

1. Archer, Gleason L. Encyclopedia of Bible Difficulties, p282-283 cites evidence based on comparing the book of Daniel to Dead Sea Scrolls texts for dating the composition of Daniel. The mismatch puts the recorded date for the book of Daniel in the late sixth century to fifth century BC period.

   In addition, the cited studies reveal the location of Daniel being written. The data supports Daniel wrote the texts at a location near Babylon.

   Finally, the evidence of the 14,000 days sequences validate Archer's viewpoint. Data is the foundation of truth. Data is the source of both the cited linguistic studies as well as my time-based studies.

   People who disagree have created a false paradigm. They base their viewpoint on opinion that denies the supernatural realm. Hence, we can logically conclude that the 165 BC composition date viewpoint is based on opinion only. Atheist assumptions are lacking. I unveil the problems with Atheism in the second and third books in the Expect to Live Forever book series.

Pre-order — Sample Chapter

1. A subatomic particle
2. Atheism is a *belief system* that denies the existence of God. In truth, Atheism is a belief system about ultimate reality. For sure, science supports that Atheism is false.
3. Website or search for the name of this scientist: https://en.wikipedia.org/wiki/Hermann_von_Helmholtz
4. Pauken, Michael, Thermodynamics for Dummies. Biography of author Michael Pauken reveals: Dr. Pauken is the senior mechanical engineer at NASA. He develops spacecraft thermal control systems and balloon systems for exploring other planets in the solar system.

Made in United States
Orlando, FL
10 November 2026

90498875R00083